Heat Transfer

R. H. S. Winterton

School of Manufacturing and Mechanical Engineering
University of Birmingham

Series sponsor: **ZENECA**

ZENECA is a major international company active in four main areas of business:
Pharmaceuticals, Agrochemicals and Seeds, Specialty Chemicals, and Biological Products.

ZENECA's skill and innovative ideas in organic chemistry and bioscience create products
and services which improve the world's health, nutrition, environment, and quality of life.
ZENECA is committed to the support of education in chemistry and chemical engineering.

OXFORD   NEW YORK   TOKYO
OXFORD UNIVERSITY PRESS
1997

Oxford University Press, Great Clarendon Street, Oxford OX2 6DP

Oxford   New York
Athens   Auckland   Bangkok   Bogota   Bombay   Buenos Aires
Calcutta   Cape Town   Dar es Salaam   Delhi   Florence   Hong Kong
Istanbul   Karachi   Kuala Lumpur   Madras   Madrid   Melbourne
Mexico City   Nairobi   Paris   Singapore   Taipei   Tokyo   Toronto
and associated companies in
Berlin   Ibadan

Oxford is a trade mark of Oxford University Press

Published in the United States
by Oxford University Press Inc., New York

A catalogue record for this book is available from the British Library

Library of Congress Cataloging in Publication Data
Winterton, R. H. S.

Heat transfer / R.H.S. Winterton.
(Oxford chemistry primers; 50)
Includes bibliographical references and index.
1. Heat Transmission. I. Title. II. Series.
QC320.W63   1997   621.402′2—dc21   97-126575
ISBN 0 19 856297 7

Typeset by EXPO Holdings, Malaysia

Printed in Great Britain by
The Bath Press

Preface

For years I have maintained that many modern textbooks are too long. My intention in writing this book has been to cover the essential material of a text twice the length. Whether I have succeeded is for others to judge.

Obviously not all heat transfer topics can be covered in book of this size. Two phase heat transfer, and advanced radiation topics, are covered in other books in this same series. Numerical techniques of solving heat transfer problems have been omitted, on the grounds that, increasingly, computer programs will be available to solve numerical problems. The emphasis in this text is on heat transfer principles, not on heat exchangers.

Birmingham   R. H. S. W.
July 1997

Series Editor's Foreword

The Oxford Chemisty Primers are designed to provide concise introductions to a wide range of topics that may be encountered by chemistry and chemical engineering students. The principles of heat transfer are central to the understanding of almost all chemical processing operations and therefore lie at the heart of any undergraduate course in chemical engineering. This Chemical Engineering Primer introduces the essential concepts in heat transfer with particular focus on conduction, forced convection, and natural convection. The theoretical arguments upon which heat transfer calculations are based are clearly explained and their use illustrated with detailed worked examples.

This Primer will be of interest to all students (and others!) who wish to obtain a practical and up-to-date understanding of heat transfer phenomena in chemical engineering.

Lynn F. Gladden
Department of Chemical Engineering, University of Cambridge

Contents

Abbreviations

Symbol		S.I. unit
A	Area, surface area (Chapters 3 and 5)	m^2
AR	aspect ratio, cavity height/width (Chapter 4)	–
a	outer radius of cylinder (Chapter 2), inner radius of tube(Chapter 3)	m
Bi	Biot number, hL/k, equation 3.40	–
c	specific heat capacity	J kg^{-1} K^{-1}
$(c$	equation 5.3 only, speed of light	m s^{-1}
C	constant defined in text	
d	diameter	m
d_e	equivalent diameter, equation 3.30 ($= d$ for circular tube)	m
D	outer diameter of tube, cylinder, or sphere	m
F_{12}	view factor	–
g	acceleration due to gravity	m s^{-2}
Gr	Grashof number, equation 4.15	–
G	irradiation	W m^{-2}
h	heat transfer coefficient, $q/(T_w - T_b)$	Wm^{-2} K^{-1}
H	enthalpy	J kg^{-1}
$(h$	equation 5.3 only, Planck's constant	J s)
J	radiosity	W m^{-2}
k	thermal conductivity	W m^{-1} K^{-1}
k_B	Boltzmann constant, equation 5.3	J K^{-1}
L	length	m
m	mass flow rate	kg s^{-1}
m	defined by equation 2.29 (Chapter 2)	
n	index various equations	–
Nu	Nusselt number, hd_e/k	–
P	pitch (distance between tube centres)	m
Pr	Prandtl number, $\mu c/k$	–
Q	heat flow rate	W
q	heat flux	W m^{-2}
r	radius	m
s	spacing of fins	m
Ra	Rayleigh number, equation 4.16	–
Re	Reynolds number, $\rho u d_e/\mu$	–
L	length; vertical length of plate	m
St	Stanton number, $h/(\rho\, uc)$	–
t	time	s
T	temperature	K
T_b	bulk or mixed mean fluid temperature	K
T_w	wall temperature (in contact with coolant)	K

T_f	film temperature, $(T_w + T_b)/2$	K
T_∞	undisturbed fluid temperature (distance from surface)	K
s	spacing (gap) between fins	m
U	overall heat transfer coefficient	W m^{-2} K^{-1}
u	heat generation rate per unit volume (Chapter 2)	W m^{-3}
v	mean velocity; local velocity (Chapter 4)	m s^{-1}
v_0	reference velocity (Chapter 4)	m s^{-1}
V	volume	m^3
x	distance	m
y	distance	m

Greek symbols

α	thermal diffusivity, $k/\rho c$, Chapter 2	m^2 s^{-1}
α	absorptivity, Chapter 5	–
β	volume coefficient of expansion	–
δ	laminar sub-layer thickness; gap width (pages 55 and 56)	m
ε	emissivity	–
θ	angle of inclination to vertical (Chapter 4);	(°)
	temperature difference, T minus reference temperature	K
λ	wavelength (Chapter 5)	m(5.3), μm (5.4)
μ	viscosity	kg m^{-1} s^{-1}
ρ	density	kg m^{-3}
ρ	reflectivity (Chapter 5)	–
σ	Stefan-Boltzmann constant, 5.67×10^{-8}	W m^{-2} K^{-4}
τ	thickness of fin	m
ϕ	angle	
ω	angular frequency $= 2\pi \times$ frequency	s^{-1}

Subscripts

av	average
b	bulk
w	wall
x	local value at position x

1 Introduction

The three basic heat transfer mechanisms, conduction, convection and radiation, are probably already familiar. In the context of heating a room by a hot water central heating system, (Fig. 1.1) all mechanisms are present. Heat is transferred into the room from the radiator both by direct **radiation** (infrared waves travelling straight out at the speed of light) and by the rising current of warm air. This last process is **natural convection,** since there is a flow of fluid and the buoyancy of the air itself causes the movement.

The hot water inside the radiator has come from the boiler under the influence of the pump, i.e. **forced convection**. If we wanted to consider the heat transfer processes from the water to the inside of the pipe, or to the inside of the radiator, then the heat transfer is forced convection. And what about conduction? Although confined to a physically small space in the figure, conduction is an essential link in the complete chain. The heat can only travel through the metal of the radiator by **conduction**.

With the two different types of convection there are in total four types of heat transfer, and these form the subject of the four main chapters of this book, starting with conduction. Unfortunately practical problems do not normally divide neatly into purely conduction problems, purely forced convection, and so on. To help overcome this the convection heat transfer coefficient is defined early on in the conduction chapter, and included in examples and problems. However, the method that has been used to calculate the convection heat transfer coefficient is not given until later.

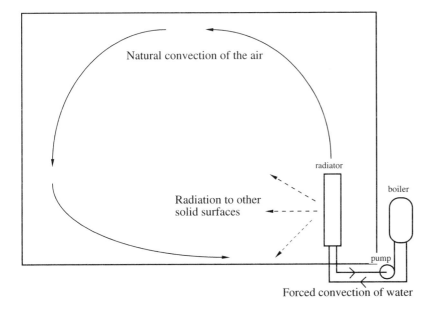

Fig. 1.1 Heat transfer mechanisms in a room

The symbols used for the various quantities are similar to those widely used in current textbooks and in research journals. The use of k for thermal conductivity and h for heat transfer coefficient goes back to Fourier in 1807. There has been a move recently to use different symbols. The problem arises in connection with thermodynamics (where h is commonly used for enthalpy). Since this is not a thermodynamics textbook we will keep to the symbols used by Fourier.

Although this text is intended to stand alone as an introduction to the subject, a couple of comments on its relation to other fields of study might be helpful. Certainly some facility with algebraic manipulation and with calculus is assumed.

A physical understanding of convection heat transfer is helped by prior study of the behaviour of the boundary layer, the thin layer of slow-moving fluid in immediate contact with the solid heat transfer surface. In many cases the boundary layer presents the main barrier to heat transfer. It does not follow, however, that thorough knowledge of fluid mechanics enables one generally to calculate convection heat transfer from first principles; empirical inputs are often needed and even in simple cases the calculation is complex. An example of what can be done starting from first principles, for a simple geometry and a simple type of flow, is given at the beginning of the natural convection chapter. The usual approach to convection is that direct measurements of heat transfer are needed as the starting point.

Another area of study that is considered to be closely related is that of thermodynamics. In particular a basic principle of heat transfer, that heat flows from a high temperature to a low temperature, is often stated to be a consequence of the second law of thermodynamics. Indeed it is, but the 2nd law is more powerful than that, since it is also concerned with the conversion of heat into work. The less general principle, that heat flows from hot to cold, was known before the 2nd law was formulated.

It should be stressed that part of the reason that the length of the book has been kept short is that alternative heat transfer coefficient correlations have generally been omitted. An advantage of having two or three correlations, all valid under the same conditions, is that they can all be applied to the same problem and the different answers compared. This is a very effective way of making the point that the error in any particular equation is probably 5 or even 10%. Since alternative methods of calculation are not, for the most part, given in this book, it is even more important to bear in mind that the error in any calculated heat transfer coefficient could be as high as 10%.

2 Conduction

2.1 Introduction

At the molecular level the internal energy of a material can be regarded simply as the kinetic energy of the atoms and molecules. Conduction heat transfer is the flow of this kinetic energy from one molecule to the next by direct contact between the two.

The Fourier law

The Fourier law states that the rate of heat flow Q is proportional to the cross-sectional area A available for heat transfer and to the temperature gradient, dT/dx, see (Fig. 2.1).

$$Q = -kA\frac{dT}{dx} \qquad \text{W} \qquad (2.1)$$

The minus sign follows from the fact that the heat must flow in the opposite direction to the temperature gradient. This point was well appreciated by Fourier[1] although it is now regarded as a consequence of the second law of thermodynamics. The constant of proportionality, k, is the **thermal conductivity**. Inspection of equation 2.1 shows that the units of thermal conductivity are $\text{W m}^{-1}\text{ K}^{-1}$.

The equation is written for the simple one-dimensional case. If temperatures are varying, and heat is flowing, in two or three directions then the equivalent of equation 2.1 may be written down for each direction. For the moment we assume one-dimensional heat flow and **steady state** conditions. Steady state means that temperatures do not change with time and there is no increase or decrease of stored energy.

Integration of equation 2.1, for cases where Q is known to be constant and the geometry simple, is straightforward. For the one-dimensional case of Fig. 2.2, where the temperature of the plane bounding one side of the slab is T_1 and that of the other side is T_2, any heat entering side 1 must leave side 2 (steady state). The same applies to the heat flow rate at any intermediate position. Q is constant and so, in this case, is A. Equation 2.1 shows that the temperature gradient is constant and

$$-\frac{Q}{kA}\int_1^2 dx = \int_1^2 dT$$

or

$$Q = -kA\frac{(T_2 - T_1)}{(x_2 - x_1)} \qquad (2.2)$$

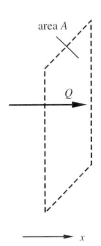

Fig. 2.1 Heat flow inside a solid

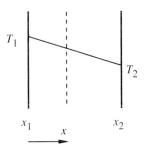

Fig. 2.2 Uniform heat flow in plane geometry

Heat flux

Often it is convenient to combine Q and A and use the **heat flux**

$$q = \frac{Q}{A} = -k\frac{dT}{dx} \qquad \text{W m}^{-2} \qquad (2.3)$$

i.e. the heat flow rate per unit area.

Values of thermal conductivity

Some typical values are shown in Fig. 2.3. The figure excludes some unusual materials such as single crystals. Also insulating materials that contain a vacuum have been omitted (in these materials the heat transfer process is no longer purely conduction). Thermal conductivity is an experimentally measured quantity; nonetheless it is interesting to try to see some pattern in the data.

The place to start, perhaps, is with non-metallic liquids. Here we can view the conduction process as being due simply to vibration in one molecule causing vibration in the next molecule and so on. This is relatively inefficient.

In solids there is the possibility of vibration waves travelling past the nearest neighbour molecules and taking vibration energy to more distant molecules directly. The atoms are arranged in fixed positions in a well-defined lattice structure and vibrate about these fixed locations. The intermolecular forces that hold the atoms in position have the effect that one atom cannot vibrate without causing vibration waves that propagate at the speed of sound. The better the long-range order in the material the more effective these waves will be, i.e. solids will be better conductors than liquids. An extreme case of long range order comes with single crystals. The thermal conductivity of diamond can reach $1000 \text{ W m}^{-1} \text{ K}^{-1}$ or more.

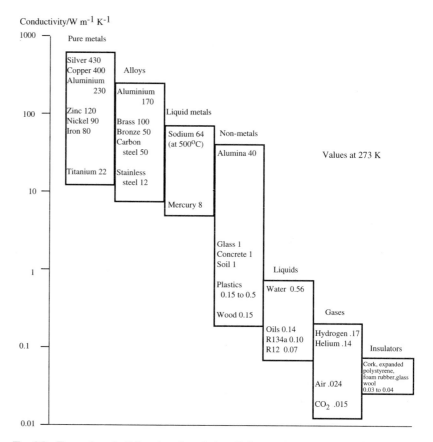

Fig. 2.3 Thermal conductivity values (mostly from Reference 2, more detailed information in Reference 3).

An additional method of passing energy directly to more distant atoms exists in metals, where free electrons carry the thermal energy. Metals have higher thermal conductivities and there is a strong correlation between thermal and electrical conductivity. Again long-range order helps the flow of electrons and so pure metals have higher k values than alloys. Frequently just a few per cent of an alloying element reduces the thermal conductivity significantly.

With gases the mechanism is different. Long-range energy transfer involving several molecules is impossible. Vibration of one molecule directly affecting a neighbour is impossible, until they collide. Thermal conductivities are low. Quite simple kinetic theory of gases can be used to give useful insight into the behaviour[4]. At low pressures there are only a few molecules to carry the heat but they can travel a long way before they collide with another molecule. These two effects cancel out giving a thermal conductivity independent of pressure (true over a wide range). The ability of the molecules to transfer energy increases with their speed, i.e. k increases with temperature. Since light molecules travel faster low molecular mass is associated with higher thermal conductivity.

To the extent that the gas is still showing much the same thermal conductivity even at very high pressures one might expect some overlap with liquid values. This is the case in Fig. 2.3. Some of the refrigerants, such as R12 (an old refrigerant being phased out) and R134a (a newer one without chlorine) have very low k values.

There is also overlap between the thermal conductivities of gases and those of insulators. This is hardly surprising since good thermal insulators are composed mainly of air. The function of the solid matrix is to keep the air still and prevent convection currents being set up.

The values of thermal conductivity show some variation with temperature. For most materials, over a modest temperature range, this variation is small. It is assumed in all the calculations in this chapter that the thermal conductivity is constant. Obviously more accurate results will be obtained if the constant value of k chosen is appropriate for the temperature range of the problem.

Example 2.1

Tests on an instrumented internal combustion engine, with thermocouples at various points inside the metal of the cylinder block, show a linear temperature variation (Fig. 2.4) with a slope of $-19\ 300$ K m^{-1}.

What is the heat flux in this region? Thermal conductivity of the cast iron is 47 W m^{-1} K^{-1}.

In deriving equation 2.2 it was assumed that the heat flux $q = Q/A$ is constant, leading to a constant temperaure gradient. We can now invert the reasoning and say that a linear temperature gradient implies a constant heat flux.

Equation 2.2 gives the heat flux as

$$q = \frac{Q}{A} = -k\frac{(T_2 - T_1)}{(x_2 - x_1)} = -47 \times (-19\ 300) = 907 \qquad \text{kW m}^{-2}$$

2.2   One-dimensional steady state conduction

The case of the plane slab has already been considered (equation 2.2). We go on to consider some more simple cases, in plane or cylindrical geometry,

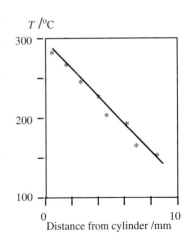

Fig. 2.4   Temperatures in an engine block

allowing for the effect of multiple layers of different materials and, in due course, of heat convection to the outside of the solid material. Again the temperature at any given point is assumed constant so there is no accumulation of stored energy.

Composite plane wall

The wall is made up of different materials with different thermal conductivities. As shown in Fig. 2.5 there are three layers, but the generalization to an indefinite number of layers is trivial. We assume good contact between the layers at points 1 and 2, i.e. there is no discontinuity in temperature at these points.

The problem (normally) is to find the heat flow rate knowing the outermost temperatures, T_0 and T_3. The intermediate temperatures are usually of less interest and can be eliminated. Since the cross-sectional area A is constant it is convenient to work in terms of $q = Q/A$. For each layer (*i*) we can use equation 2.2:

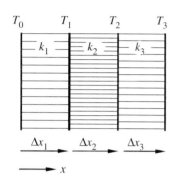

Fig. 2.5   Plane wall made of 3 different materials

$$q = -k_i \frac{(T_i - T_{i-1})}{\Delta x_i}$$

or
$$(T_i - T_{i-1}) = -q \frac{\Delta x_i}{k_i}$$

writing this down for all three layers:

$$T_3 - T_2 = -q \frac{\Delta x_3}{k_3} \tag{2.4}$$

$$T_2 - T_1 = -q \frac{\Delta x_2}{k_2} \tag{2.5}$$

$$T_1 - T_0 = -q \frac{\Delta x_1}{k_1} \tag{2.6}$$

Adding the three equations gives

$$T_3 - T_0 = -q \left(\frac{\Delta x_1}{k_1} + \frac{\Delta x_2}{k_2} + \frac{\Delta x_3}{k_3} \right)$$

Making q the subject and generalising for n layers gives

$$q = -\frac{(T_n - T_0)}{\sum\limits_{1}^{n} \frac{\Delta x_i}{k_i}} \tag{2.7}$$

Composite plane wall with heat transfer to fluid on each side

Unfortunately it is not often that the temperature of the immediate surface of the wall is known. If Fig. 2.5 is considered to be the outside wall of a house then the interior temperature of the room might be known, say 20 °C. Also the temperature of the air outside the house might be known, say 10 °C. To make the problem more realistic we need to introduce a parameter that enables us to calculate what happens between a fluid in contact with a wall and the wall itself. This parameter will not be fully explained until the next chapter.

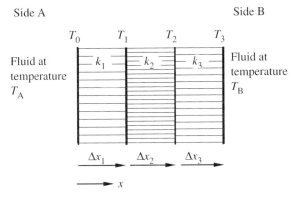

Side A   Side B

T_0 T_1 T_2 T_3

Fluid at k_1 k_2 k_3 Fluid at
temperature   temperature
T_A T_B

Δx_1 Δx_2 Δx_3

x

Fig. 2.6 Composite wall with heat transfer to fluid on each side

For now it is enough to know that the convective heat flux is given by

$$q = h(T_w - T_b) \qquad \text{W m}^{-2} \tag{2.8}$$

or
$$Q = hA(T_w - T_b) \qquad \text{W}$$

where h is the **heat transfer coefficient,** with units of W m^{-2} K^{-1}.

T_w is the temperature of the wall and T_b is the temperature of the bulk fluid some distance away from the wall. In the case of the example of the house wall T_b inside is the room air temperature, 20 °C and T_b outside is the ambient air temperature, 10 °C. The heat transfer coefficient h could well be different inside and outside. At the moment we have no method of calculating h; the value has to be given.

If $T_w > T_b$ then, fairly obviously, heat flow is from the wall into the fluid. With walls facing in different directions we will need to be consistent in the sign of Q.

We are now able to analyse the problem (Fig. 2.6), including consideration of the fluid on each side. The bulk fluid temperature on side A is T_A and that on side B is T_B. The bulk fluid temperature is often written T_∞ which emphasizes the point that this is the temperature of the fluid far from the wall.

Starting at side B, using equation 2.8:

$$q = h(T_3 - T_B)$$

or
$$T_B - T_3 = -\frac{q}{h_B} \tag{2.9}$$

where h_B is the heat transfer coefficient on side B. There is no need to write down the equations for the conduction steps inside the wall—they are the same as before, equations 2.4 to 2.6.

On side A we have

$$T_0 - T_A = -\frac{q}{h_A} \tag{2.10}$$

(the minus sign follows since heat flow will be in the negative x direction for $T_0 > T_A$.)

Adding all the equations, i.e. 2.9, 2.4 to 2.6, and 2.10, gives

$$T_B - T_A = -q\left\{\frac{1}{h_B} + \sum\frac{\Delta x_i}{k_i} + \frac{1}{h_A}\right\}$$

or

$$q = -\frac{T_B - T_A}{\left\{\dfrac{1}{h_B} + \sum\dfrac{\Delta x_i}{k_i} + \dfrac{1}{h_A}\right\}} \qquad (2.11)$$

where the result has been generalized to a wall with an indefinite number of parallel layers.

Comparing equation 2.11 with the original equation, 2.8, that defined the heat transfer coefficient, we could write

$$q = -U(T_B - T_A) \qquad (2.12)$$

Where U is the **overall heat transfer coefficient**.

$$\frac{1}{U} = \frac{1}{h_B} + \sum\frac{\Delta x_i}{k_i} + \frac{1}{h_A} \qquad (2.13)$$

Example 2.2
A double-glazed window 1.2 m high by 2 m wide separates a room at 22 °C from the outside of the building at –2 °C (Fig. 2.7). The panes are 5 mm thick and the gap between them is 12 mm. The thermal conductivity of the glass is 0.8 W m^{-1} K^{-1} and the heat transfer coefficient is the same on each side at 2.3 W m^{-2} K^{-1}. Assume pure conduction for the air in the gap. What is the rate of heat loss through the window?

The problem is symmetrical so the temperature of the air in the gap will be 10 °C. From the table in the Appendix the thermal conductivity is (0.024 11 0.0256)/2 = 0.024 85 Wm^{-1} K^{-1}. Substituting all these values in equation 2.11 the heat flux through the window is

$$q = -\frac{T_B - T_A}{\left\{\dfrac{1}{h_B} + \sum\dfrac{\Delta x_i}{k_i} + \dfrac{1}{h_A}\right\}} = -\frac{-2 - 22}{\left\{\dfrac{1}{2.3} + \dfrac{.005}{.8} + \dfrac{.012}{.02485} + \dfrac{.005}{8} + \dfrac{1}{2.3}\right\}}$$

$$= 17.6 \quad \text{W m}^{-2}$$

and the total rate at which heat is lost is $1.2 \times 2 \times 17.6 = 42.2$ W

In Chapter 4 (example 4.2) the method of calculating the heat transfer coefficients is explained and the assumption of pure conduction in the air in the gap justified (to within a few per cent anyway).

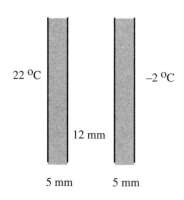

22 °C –2 °C

12 mm

5 mm   5 mm

Fig. 2.7 Double glazed window of example 2.2

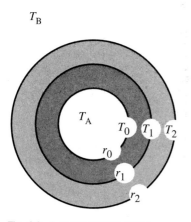

T_B

T_A
T_0 T_1 T_2
r_0
r_1
r_2

Fig. 2.8 Lagged cylindrical pipe

Cylindrical tube with heat transfer to fluid inside and out

A cross-section of the geometry is shown in Fig. 2.8. It is assumed that the pipe consists of more than one layer. It could be a lagged metal pipe containing steam. The bulk temperature of the fluid on the inside is T_A and that of the fluid on the outside is T_B.

The problem of finding the heat flow in terms of T_A and T_B, which may be the only known temperatures, is similar to that for the plane wall except that we cannot assume that the heat flux q is constant. With the cylindrical geometry only the total heat flow rate Q is constant (assuming again steady state conditions). The heat flux will in fact decrease as the heat flows outwards.

In cylindrical co-ordinates the Fourier equation becomes

$$Q = -kA\frac{\mathrm{d}T}{\mathrm{d}r} = -k2\pi rL\frac{\mathrm{d}T}{\mathrm{d}r}$$

where L is the length of the pipe (assumed sufficiently long for essentially one-dimensional heat flow). Within a given region, say between r_0 and r_1:

$$\int_{r_0}^{r_1} \frac{Q}{2\pi k_1 L}\frac{\mathrm{d}r}{r} = -\int_{T_0}^{T_1} \mathrm{d}T$$

or

$$\frac{Q}{2\pi k_1 L}\ln\frac{r_1}{r_0} = T_0 - T_1 \qquad (2.14)$$

Writing the temperature differences down for each region, taking heat flow out (in the positive r direction) as positive

$$T_\mathrm{A} - T_0 = \frac{q_0}{h_\mathrm{A}} = \frac{Q}{2\pi r_0 L}\frac{1}{h_\mathrm{A}} \qquad (2.15)$$

$$T_0 - T_1 = \frac{Q}{2\pi k_1 L}\ln\frac{r_1}{r_0} \qquad (2.16)$$

$$T_1 - T_2 = \frac{Q}{2\pi k_2 L}\ln\frac{r_2}{r_1} \qquad (2.17)$$

$$T_2 - T_\mathrm{B} = \frac{q_2}{h_\mathrm{B}} = \frac{Q}{2\pi r_2 L}\frac{1}{h_\mathrm{B}} \qquad (2.18)$$

Adding equations 2.15 to 2.18:

$$T_\mathrm{A} - T_\mathrm{B} = \frac{Q}{2\pi L}\left\{\frac{1}{r_0 h_\mathrm{A}} + \frac{1}{k_1}\ln\frac{r_1}{r_0} + \frac{1}{k_2}\ln\frac{r_2}{r_1} + \frac{1}{r_2 h_\mathrm{B}}\right\} \qquad (2.19)$$

and with more solid layers the result could easily be generalized.

Heat conduction from buried cylinders and spheres

A topic of interest in more than one connection is the total amount of heat lost from a long cylinder or from a sphere which is surrounded by conducting material. One application is simply the heat transfer from objects buried in the ground. Another is conduction from objects surrounded by a fluid. Although in this case convective heat transfer may be more important, the conduction into the surrounding fluid represents a lower limit to the total heat transfer, still present when the convective heat transfer becomes negligibly small.

Cylinder
We have already integrated the Fourier equation for the cylindrical geometry, giving equation 2.14. Since there is now only the one material we can replace k_1 by k. Also the result is valid for the temperature at any radius r so we replace r_1 by r:

$$\frac{Q}{2\pi kL}\ln\frac{r}{r_0} = T_0 - T_r$$

The cylinder is assumed isothermal with surface temperature T_0. For a cylinder buried in an indefinite extent of material r tends to infinity, as does $\ln r$, which means Q/L tends to zero. This is not a particularly interesting

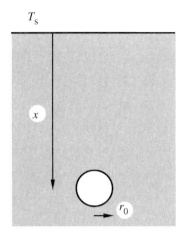

T_S

x

r_0

Fig. 2.9   Buried cylinder

result, because strictly it applies only to an infinitely long cylinder, but we refer to it later in the convection chapters.

If the material does not extend to infinity, but the centre of the cylinder is buried a distance x below the surface (Fig. 2.9), which is isothermal at temperature T_S, then it can be shown that

$$Q = k(T_0 - T_S)\frac{2\pi L}{\cosh^{-1}\dfrac{x}{r_0}} \tag{2.20}$$

provided $L \gg r_0$.

Sphere

For a sphere, radius r_0, surrounded by a spherical layer of material, the Fourier equation becomes

$$Q = -kA\frac{\mathrm{d}T}{\mathrm{d}r} = -k4\pi r^2\frac{\mathrm{d}T}{\mathrm{d}r}$$

so

$$\int_{r_0}^{r}\frac{Q}{4\pi k}\frac{\mathrm{d}r}{r^2} = -\int_{T_0}^{T_r}\mathrm{d}T$$

or

$$Q\left\{\frac{1}{r_0} - \frac{1}{r}\right\} = 4\pi k(T_0 - T_r) \tag{2.21}$$

and as r tends to infinity this becomes

$$Q = 4\pi k r_0(T_0 - T_\infty)$$

In terms of heat flux on the surface of the sphere this is

$$q = \frac{k}{r_0}(T_0 - T_\infty) \tag{2.22}$$

Alternatively this equation could be put in the form of a dimensionless heat transfer coefficient using the diameter of the sphere, $2r_0$. Since $q = h(T_0 - T_\infty)$ we have $h = k/r_0$ and

$$\frac{h2r_0}{k} = 2$$

This result is widely used as the lower limit in convective heat transfer correlations for spheres. Results for many other buried geometries are given in Reference 5.

Temperature distribution in a cylinder with heat generation

Several heat conduction problems involve heat generation within the bulk of the material. Two examples are the heat produced by fission in the uranium of a nuclear reactor fuel element and the Joule heating of an electric current passing along a wire. Both of these cases commonly involve a cylindrical geometry.

To make the problem more complicated a hole has been added in the centre of the cylinder (Fig. 2.10). This is not unusual in the case of uranium dioxide fuel pellets.

The outer radius of the cylinder is a and that of the hole r_h. The heat appears at a uniform rate u per unit volume. The parameter of greatest interest is often the maximum temperature, since there may be materials limitations on the maximum permitted temperature.

Consider a cylinder of length unity, radius r, concentric with the main cylinder. The rate of heat production within the radius r is

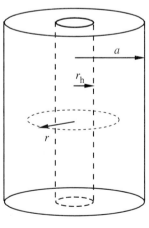

a

r_h

r

Fig. 2.10   A hollow cylinder with heat generation

$$u\pi(r^2 - r_{\mathrm{h}}^2)$$

since the power produced is in proportion to the volume of material.

Now in equilibrium (i.e. in the steady state) this rate of internal energy production must equal the rate at which heat is conducted out through the cylindrical surface at r, which by the Fourier law is

$$-k2\pi r\frac{\mathrm{d}T}{\mathrm{d}r}$$

Equating these and integrating gives

$$\frac{u}{2k}\int_r^a \left(r - \frac{r_{\mathrm{h}}^2}{r}\right)\mathrm{d}r = -\int_{T_r}^{T_a} \mathrm{d}T$$

So the temperature at radius r is

$$T_r = T_a + \frac{u}{2k}\left[\frac{a^2 - r^2}{2} - r_{\mathrm{h}}^2\ln\frac{a}{r}\right] \tag{2.23}$$

and the maximum temperature in the material is found by putting $r = r_{\mathrm{h}}$.

For a solid material $r_{\mathrm{h}} = 0$ and

$$T_r = T_a + \frac{u}{4k}\left[a^2 - r^2\right] \tag{2.24}$$

or
$$T_{\max} = T_a + \frac{ua^2}{4k} \tag{2.25}$$

To complete the analysis, in the case of a fuel rod in a nuclear reactor, one would need to consider the heat flow through the cylindrical can (equation 2.14) and the possiblitity of contact resistance at the interface between the metal can and the ceramic fuel pellets (see later).

Improvement of heat transfer using fins

This topic involves both steady state conduction and convection. The normal approach is to assume that the convective heat transfer coefficient is constant, regardless of changes that are being made to the shape of the surface. The analysis thus is mainly concerned with conduction.

The basic idea is to increase the surface area exposed to the coolant by means of projections from the solid surface. This is often done if the surface is made of metal and the coolant is a gas. An example is shown in Fig. 2.11 where longitudinal fins have been added to the outside of a tube. From the definition of heat transfer coefficient, equation 2.8, the heat lost from the original, unfinned, surface is

$$Q_{\mathrm{plain}} = hA_{\mathrm{plain}}(T_{\mathrm{w}} - T_{\mathrm{b}})$$

and that from the finned surface is

$$Q_{\mathrm{finned}} = hA_{\mathrm{finned}}(T_{\mathrm{w}} - T_{\mathrm{b}})$$

and one might hope that $Q_{\mathrm{finned}}/Q_{\mathrm{plain}} = A_{\mathrm{finned}}/A_{\mathrm{plain}}$ We have already assumed that h is unchanged. However the assumption that the temperature of the surface of the solid, including the fins, is unchanged at T_{w}, may not be justified. Heat has to be conducted down the fins to maintain their temperature. If the thermal conductivity k of the solid material is low (or h is high) then the temperature of the fins will fall below T_{w} and the performance of the finned

Fig. 2.11 Cross-section through typical finned surface—a tube with longitudinal fins added

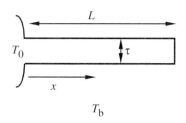

Fig. 2.12   Constant cross-section fin. Unit width perpendicular to figure is assumed

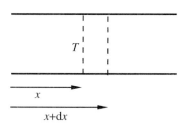

Fig. 2.13   Heat balance on an element dx of fin

surface will be worse than expected. The conduction process down the fin needs to be analysed.

Fins can be of various shapes. The simplest type to analyse is one where the cross-section is constant. We assume a fin of length L and constant thickness τ, Fig. 2.12. (It could be one of the fins in Fig. 2.11.) For simplicity consider unit width of the fin perpendicular to the plane of Fig. 2.12, i.e. the cross-sectional area of the fin for heat conduction is just τ.

Since $\tau \ll L$ it is reasonable to assume that the temperature of the fin is constant at any particular distance x along the fin; the problem is one dimensional with temperature T only varying with x. Considering the heat balance on an element of fin length dx at x (Fig. 2.13), the Fourier law gives the heat conducted in at x as

$$-k\tau \frac{dT}{dx} \qquad (2.26)$$

and the heat conducted out at $x + dx$ is at a rate

$$-k\tau \left\{ \frac{dT}{dx} + \frac{d^2T}{dx^2} dx \right\} \qquad (2.27)$$

So the net rate at which heat is being conducted into the element dx is, subtracting 2.27 from 2.26,

$$k\tau \frac{d^2T}{dx^2} dx$$

This equals the heat lost from the upper and lower surfaces of the fin

$$2 dx h (T - T_b)$$

(since T at any particular position along the fin $= T_w =$ temperature of the surface).
Equating the last two expressions and putting $\theta = T - T_b$ gives

$$\frac{d^2\theta}{dx^2} = \frac{2h}{k\tau} \theta \qquad (2.28)$$

This type of differential equation is satisfied by solutions of the form $\theta = Be^{mx}$ and substituting back into equation 2.28 shows that

$$m^2 = \frac{2h}{k\tau} \qquad (2.29)$$

i.e. m can be either positive or negative. Regarding m as positive then a general solution is

$$\theta = Be^{mx} + Ce^{-mx} \qquad (2.30)$$

The values of the constants B and C come from the boundary conditions. At $x = 0$, $\theta = \theta_0 = T_0 - T_b$ so

$$\theta_0 = B + C \qquad (2.31)$$

The second boundary condition is more difficult. If the fin was very long one could simply say that an exponentially increasing temperature term like Be^{mx} was not acceptable and therefore $B = 0$. This is a correct conclusion but not very useful: there is no point in making fins so long that the heat cannot get to the end of them. A boundary condition of zero heat flow, i.e. zero temperature gradient, at the tip of the fin would nonetheless be an easy one to apply.

This boundary condition can be achieved with a minor alteration to the problem. The end of the fin is imagined to be extended slightly, to a new total length L' with $\tau = 2(L' - L)$, and the new tip is considered to be insulated, as in Fig. 2.14. Exactly the same surface area is exposed to the coolant but we now have the simple boundary condition of no heat flow at $x = L'$.

So at $x = L$

$$\frac{d\theta}{dx} = Bme^{mL'} - Cme^{-mL'} = 0$$

substituting for C from equation 2.31 gives

$$B = \theta_0 e^{-mL'} / (e^{mL'} + e^{-mL'})$$

and hence

$$C = \theta_0 e^{mL'} / (e^{mL'} + e^{-mL'})$$

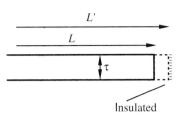

Fig. 2.14 Simplified boundary condition. Extended fin with insulated tip

Using the last two equations with equation 2.30 gives the temperature distribution along the fin, but what we are mainly interested in is the total heat dissipated by the fin, which equals the heat conducted in at the root

$$-k\tau\left(\frac{d\theta}{dx}\right)_{x=0} = -k\tau(Bm - Cm) = k\tau m\theta_0 \frac{e^{mL'} - e^{-mL'}}{e^{mL'} + e^{-mL'}}$$

where the last part of the expression is the definition of the hyperbolic tangent.
So the total rate at which heat is dissipated by the fin is

$$k\tau m\theta_0 \tanh mL' \qquad (2.32)$$

A measure of fin performance is the **fin efficiency** defined as the ratio of the heat actually dissipated by the fin to the heat that would be conveyed to the coolant if the whole surface were at the root temperature T_0. In other words the actual behaviour is compared with that of an idealized fin with infinite thermal conductivity.

If the whole surface were at T_0, the heat convected away would be

$$2L'h(T_0 - T_b) = 2hL'\theta_0$$

So the fin efficiency is

$$E = \frac{\tanh mL'}{mL'} \qquad (2.33)$$

where m is given by equation 2.29, i.e. $mL' = (2hL'^2/k\tau)^{\frac{1}{2}}$ with the extended length L' given by $\tau = 2(L' - L)$.

An indication of the variation of the fin efficiency with length is shown in Fig. 2.15. For high fin efficiency a low value of m is required and in practice

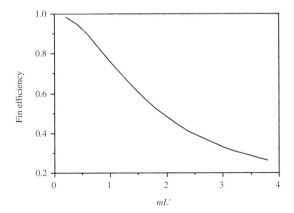

Fig. 2.15 Fin efficiency related to length of fin.

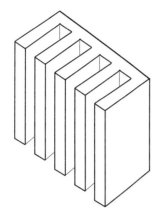

Fig. 2.16   Heat sink

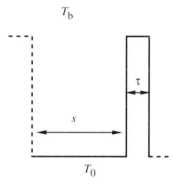

Fig. 2.17   Unit cell of finned
surface—one fin plus gap

Fig. 2.18   Contact between solid
surfaces

this normally involves two conditions:
a) a high value of k, i.e. metal fins;
b) a low value of h, i.e. cooling by air or other gas.

A small-scale example, e.g. Fig. 2.16, would be a heat sink for electronic components. The component is fixed to the other side of the heat sink and the heat it generates spread through the fins to allow heat removal without the component getting too hot. Typically the heat sink would be make of aluminium—easy to extrude into complex shapes and with a high thermal conductivity.

Total heat lost by the finned surface
The rate of heat loss by the fin is the fin efficiency E times the heat lost by a perfect fin with no temperature drop along it. The original purpose of adding fins was to extend the heat transfer surface area. If a unit cell can be identified, Fig. 2.17, such that the complete surface can be made up by replicating this basic unit, and again assuming 1 m extent of surface perpendicular to the figure, then:

$$Q_{plain} = \text{heat loss rate without fins} = \text{heat flux} \times \text{area} = h(T_0 - T_b)(s + \tau)$$

$$Q_{finned} = \text{heat loss rate with fins} = h(T_0 - T_b)s + Eh(T_0 - T_b)(2L + \tau)$$

where s is the spacing of the fins (i.e. $s + \tau$ is the pitch).

In practice it will be necessary for the ratio Q_{finned}/Q_{plain} to be significantly greater than 1. The method of calculation overstates the benefit of fins. h is likely to fall at the root of the fins where the coolant cannot penetrate. An estimate of the minimum fin spacing needed for natural convection in air is given in Chapter 4.

Contact resistance

When two solid surfaces are pressed into contact they will only touch at isolated points (Fig. 2.18). At these points of contact direct solid-to-solid conduction can occur but elsewhere heat has to be conducted through the air in the gaps (assuming the system is in air). This gives rise to a contact resistance, often expressed as a gap heat transfer coefficient, h_g.

$$q = h_g(T_1 - T_2)$$

where T_1 and T_2 are the temperatures on either side of the gap. To get a rough idea of the order of magnitude of h_g we could suppose that the gap is 10 μm wide and filled with air at 100 °C (10 μm would be considered quite a rough surface finish but there is also the question of how flat and true the surfaces are). Conduction through the air (equation 2.2) gives

$$q = (k/\Delta x)(T_1 - T_2) = h_g(T_1 - T_2)$$

or $$h_g = k/\Delta x = 0.031\ 31/10^{-5} = 3000 \qquad \text{W m}^{-2}\text{ K}^{-1}$$

Measurements of the gap heat transfer coefficient for various metal surfaces in air[5] give values varying from 2000 to 15 000 W m^{-2} K^{-1}. There is some tendency for the value to increase as the pressure pushing the surfaces together increases, but normally the increase is less than a factor of two even for a factor of 50 increase in pressure (results for 4×10^4 to 200×10^4 N m^{-2}). The significance of the conduction through the air is seen from the fact that much lower h_g values are obtained if the system is evacuated.

2.3   The general differential equation for heat conduction

The cases solved up to now have been straightforward: they have been steady state in simple one-dimensional geometries. If these assumptions are not justified then a more basic approach is needed, setting up the general differential equation that always applies and then trying to solve it using the boundary conditions of the particular problem.

Consider an elemental box $dx \times dy \times dz$ inside the material, as shown in Fig. 2.19. The rate of heat flow into the box at x, from the Fourier law, is

$$-kA\frac{\partial T}{\partial x} = -k\,dy\,dz\,\frac{\partial T}{\partial x}$$

and the rate of heat flow out of the box at $x + dx$ is

$$-k\,dy\,dz\left\{\frac{\partial T}{\partial x} + \frac{\partial^2 T}{\partial x^2}\,dx\right\}$$

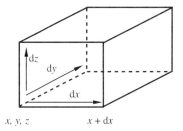

Partial derivatives, e.g. $\partial T/\partial x$, have been used to emphasize that this is just the variation with x, while y and z (and later t) are kept constant.

So the net rate of heat flow into the box in the x direction is

Fig. 2.19   Elementary box for analysis of heat flows and rate of temperature rise

$$k\,dy\,dz\,\frac{\partial^2 T}{\partial x^2}\,dx$$

Considering all three directions, the net rate of heat flow into the box is

$$k\,dx\,dy\,dz\left\{\frac{\partial^2 T}{\partial x^2} + \frac{\partial^2 T}{\partial y^2} + \frac{\partial^2 T}{\partial z^2}\right\}$$

and we can save a bit of space and ink by writing

$$\nabla^2 T = \frac{\partial^2 T}{\partial x^2} + \frac{\partial^2 T}{\partial y^2} + \frac{\partial^2 T}{\partial z^2} \tag{2.34}$$

(∇ is sometimes called **del**, i.e. an upside down capital delta.)

If heat is being generated at a rate u W m^{-3} then the rate of heat production within the box is $u\,dx\,dy\,dz$.

The heat being supplied to the box causes its temperature to rise. The definition of specific heat capacity c is the heat required to raise the temperature of 1 kg of the substance by 1 K. It follows that the thermal capacity of the elemental box, i.e. the heat required to cause a 1 K temperature rise, is $\rho c\,dx\,dy\,dz$.

So the net heat supplied in time ∂t causes a temperature rise ∂T or

$$(k\,dx\,dy\,dz\nabla^2 T + u\,dx\,dy\,dx)\partial t = \rho c\,dx\,dy\,dz\partial T$$

giving
$$\nabla^2 T + \frac{u}{k} = \frac{\rho c}{k}\frac{\partial T}{\partial t} \tag{2.35}$$

Writing $k/\rho c = \alpha$, the **thermal diffusivity**, then

$$\nabla^2 T + \frac{u}{k} = \frac{1}{\alpha}\frac{\partial T}{\partial t} \tag{2.36}$$

Special cases

Steady state

$$\nabla^2 T + \frac{u}{k} = 0 \tag{2.37}$$

No heat generation

$$\nabla^2 T = \frac{1}{\alpha} \frac{\partial T}{\partial t} \tag{2.38}$$

Steady state and no heat generation $\nabla^2 T = 0$ \qquad (2.39)

2.4   Transient problems

The equations above give the basis for solving transient, i.e. time-varying, heat conduction problems. Before starting on that it is important to realize that many problems of practical interest do not depend purely on conduction. More generally, (Fig. 2.20), the boundary condition at the surface of the solid object is likely to be one of convective heat transfer, with a heat transfer coefficient h, rather than a known temperature.

Depending on the relative efficiencies of conduction and convection three possibilities can be recognized:

h very high

This might be the case with a liquid metal or boiling coolant, or simply with a very high velocity liquid coolant. The temperature of the surface of the solid is essentially the same as that of the coolant and the boundary condition is that of constant known temperature at the surface of the solid.

h very low

This might apply for air or gas cooling of a metal. The rate of cooling is controlled by the rate at which the gas removes heat from the solid surface. There is plenty of time for temperatures within the metal to equalize so the temperature throughout the metal can be regarded as uniform at a given time. This is not a transient conduction problem and the lumped parameter solution (see next chapter) is appropriate.

h intermediate

Analysis becomes complex. A number of analytical solutions are given in Reference 6. In many cases only numerical solution by computer is possible.

Significance of thermal diffusivity

The units of α, worked out either from its definition or from equation 2.36, are $m^2\ s^{-1}$. This has a very simple interpretation. Very roughly heat will diffuse by conduction a distance x in a time t where $x^2/t = \alpha$.

One example of transient conduction where this concept is very helpful follows in the next section.

Step change of temperature at the surface of a semi-infinite solid (Fig. 2.21)

Initially at $t = 0$ the temperature is T_0 everywhere. For $t > 0$ the surface is suddenly exposed to a new temperature T_1 (i.e. an example of the $h = \infty$, or constant temperature, boundary condition). How does the temperature within the material vary with time and depth?

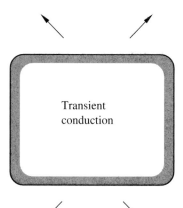

Transient
conduction

Convection (or
radiation) from
surface with heat
transfer coefficient h

Fig. 2.20   The general transient heat transfer problem

x

Fig. 2.21   Variation of temperature with time and depth x into solid

The boundary condition on the surface, at $x = 0$, is shown in Fig. 2.22. There is no heat generation so the starting differential equation is 2.38:

$$\nabla^2 T = \frac{1}{\alpha} \frac{\partial T}{\partial t}$$

and since there is only one distance variable, the depth x, the problem is one dimensional:

$$\frac{\partial^2 T}{\partial x^2} = \frac{1}{\alpha} \frac{\partial T}{\partial t} \tag{2.40}$$

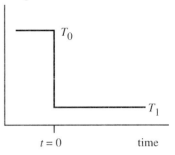

Fig. 2.22 Variation of temperature at the surface

At this point we use our physical insight into the meaning of the thermal diffusivity and assume that x and t only appear in the solution in the group $x^2/\alpha t$, i.e. we define a new dimensionless variable

$$z = \frac{x}{\sqrt{\alpha t}}$$

(The reason that this works is that x is the only distance that appears in the problem and t the only time; z, or some function of it, is the only dimensionless group that can be formed from x and t.)

We make T dimensionless by writing

$$\theta = \frac{T - T_1}{T_0 - T_1} = \frac{T - T_1}{\Delta T}$$

In other words we are expecting a dimensionless solution

$$\theta = f(z)$$

Expressions are needed for the two sides of equation 2.40 in terms of the new variable z.

$$\frac{\partial T}{\partial t} = \Delta T \frac{\partial \theta}{\partial t} = \Delta T \frac{\partial \theta}{\partial z} \frac{\partial z}{\partial t}$$

and since

$$\frac{\partial z}{\partial t} = -\frac{x}{\sqrt{\alpha}} \frac{1}{2} t^{-3/2}$$

$$\frac{\partial T}{\partial t} = -\Delta T \frac{\partial \theta}{\partial z} \frac{x}{\sqrt{\alpha}} \frac{1}{2} t^{-3/2}$$

Similarly the left-hand side of equation 2.40 is needed, starting with

$$\frac{\partial T}{\partial x} = \Delta T \frac{\partial \theta}{\partial x} = \Delta T \frac{\partial \theta}{\partial z} \frac{\partial z}{\partial x} = \Delta T \frac{\partial \theta}{\partial z} \frac{1}{\sqrt{\alpha t}}$$

so

$$\frac{\partial^2 T}{\partial x^2} = \frac{\partial z}{\partial x} \frac{\partial}{\partial z} \left[\Delta T \frac{\partial \theta}{\partial z} \frac{1}{\sqrt{\alpha t}} \right] = \Delta T \frac{1}{\alpha t} \frac{\partial^2 \theta}{\partial z^2}$$

substituting back into equation 2.40 and simplifying gives

$$\frac{\partial^2 \theta}{\partial z^2} = -\frac{z}{2} \frac{\partial \theta}{\partial z}$$

which can be written

$$\frac{d^2 \theta}{dz^2} = -\frac{z}{2} \frac{d\theta}{dz} \tag{2.41}$$

since there is now just the one variable z, replacing the two previous variables x and t.

So we have succeeded in putting the differential equation in the form $\theta = f(z)$. We now need to check that the boundary conditions can be put in terms of z.

$$T = T_0 \text{ at } t = 0 \text{ becomes } \theta = 1 \text{ at } z = \infty \tag{2.42}$$

$$T = T_1 \text{ for } x = 0 \text{ at } t > 0 \text{ becomes } \theta = 0 \text{ at } z = 0 \tag{2.43}$$

These too are solely functions of z. Mathematically speaking we now have a much simpler equation to solve (equation 2.41) compared to the one we started with (equation 2.40).

The solution proceeds in two stages, starting with $d\theta/dz$. Writing

$$d\theta/dz = \phi$$

i.e. equation 2.41 becomes

$$\frac{d\phi}{dz} = -\frac{1}{2} z\phi$$

or

$$\frac{d\phi}{\phi} = -\frac{1}{2} z dz$$

which integrates to $\ln\phi = -\frac{1}{4}z^2 + C$ where C is a constant.

So

$$\phi = \frac{d\theta}{dz} = C e^{-z^2/4}$$

and we can start on the next stage

$$[\theta]_0^\theta = C \int_0^z e^{-z^2/4} dz$$

or

$$\theta = C \int_0^z e^{-z^2/4} dz \qquad \text{since } \theta \text{ at } z = 0 \text{ is } 0.$$

The other boundary condition, at $z = \infty$, gives

$$1 = C \int_0^\infty e^{-z^2/4} dz$$

and tables of integrals reveal that C is $1/\sqrt{\pi}$ and that

$$\frac{1}{\sqrt{\pi}} \int_0^z e^{-z^2/4} dz$$

is called the error function of $z/2$, erf $z/2$.

So the solution is

$$\theta = \operatorname{erf} \frac{z}{2} \tag{2.44}$$

An indication of the behaviour of this function is shown in Fig. 2.23. Values of the function are tabulated in the Appendix.

The analysis of these transient problems, as we have seen above, is often quite lengthy. For the next two problems the result is given but not the proof. Many other results are available in Reference 6.

Sinusoidally varying temperature at the surface of a semi-infinite solid

This could, for example, be a model for the daily variation of temperature at the surface of the earth. We are concerned with variations above and below the

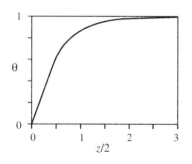

Fig. 2.23 Dimensionless temperature versus dimensionless distance/time

average temperature T_{av}. At the surface

$$T_{x=0} = T_{av} + \Delta T \cos\omega t$$

where ω is the angular frequency of the variation. The period of the oscillation is $2\pi/\omega$, so $1/\omega$ is a measure of the timescale. It is assumed that this sinusoidal variation has existed for a long time and so any initial transients can be ignored.

As regards the variation with depth x into the surface, since x is the only linear dimension that appears in the problem, we can expect that x will only appear in the group $x\sqrt{\omega/\alpha}$ (from the physical interpretation of the thermal diffusivity, α). It is to be expected that the oscillations in temperature inside the slab will still have the angular frequency ω but will lag behind those on the surface. Also the oscillations inside the slab will be diminished in amplitude.

It can be shown that the solution is

$$\frac{T - T_{av}}{\Delta T} = e^{-x\sqrt{\omega/2\alpha}} \cos\left(\omega t - x\sqrt{\frac{\omega}{2\alpha}}\right) \qquad (2.45)$$

Infinite plate with step change of surface temperature

The initial temperature of the plate is uniform at T_0 (Fig. 2.24). The surface is suddenly exposed to a new temperature T_1. How does the temperature inside the plate vary with depth and time? Since the width of the plate is finite at L we can expect that x/L will appear in the solution as well as $L/\sqrt{\alpha t}$. It can be shown that the solution is in the form of an infinite series

$$\frac{T - T_1}{T_0 - T_1} = \frac{4}{\pi} \sum_{n=1}^{\infty} \frac{1}{n} e^{-[n\pi/L]^2 \alpha t} \sin\frac{n\pi x}{L} \qquad (2.46)$$

where n takes odd values only, i.e. $n = 1, 3, 5$, etc.

It is interesting to note that the solution of this problem involves Fourier series, something that Fourier was already working on in Reference 1 and is perhaps more famous for.

Three-dimensional problems

Some three-dimensional problems can be solved simply by combining solutions to the equivalent one-dimensional problem. For transient problems without heat generation the differential equation to be solved is (2.38)

$$\frac{\partial^2 T}{\partial x^2} + \frac{\partial^2 T}{\partial y^2} + \frac{\partial^2 T}{\partial z^2} = \frac{1}{\alpha}\frac{\partial T}{\partial t} \qquad (2.38)$$

Suppose that the one-dimensional problem

$$\frac{\partial^2 T}{\partial x^2} = \frac{1}{\alpha}\frac{\partial T}{\partial t}$$

has been solved and the solution in terms of dimensionless temperature θ is

$$\theta = X(x, t)$$

where X is a function of just x and t (e.g. equations 2.45 and 2.46). Similarly Y ($= Y(y, t)$) and $Z(= Z(z, t)$) are the solutions to the one-dimensional problem in the y and z directions. In the simplest case X, Y and Z could all have the

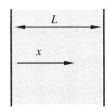

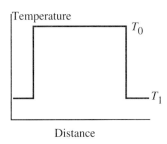

Fig. 2.24   Temperature of the plate at time 0

same mathematical form. It can be shown[6] that the solution to the three-dimensional problem is

$$\theta = XYZ$$

It is still necessary to check on the boundary conditions.

For the specific case of a rectangular box, $0 < x < x_1$, $0 < y < y_1$, and $0 < z < z_1$, initially all at temperature T_0 but with the surface suddenly exposed to a temperature T_1, the one-dimensional solution was equation 2.46. So the three-dimensional solution is

$$\frac{T - T_1}{T_0 - T_1} = \theta = XYZ$$

$$= \frac{64}{\pi^3} \left\{ \sum_{n=1}^{\infty} \frac{1}{n} e^{-[n\pi/x_1]^2 \alpha t} \sin \frac{n\pi x}{x_1} \right\} \left\{ \sum_{n=1}^{\infty} \frac{1}{n} e^{-[n\pi/y_1]^2 \alpha t} \sin \frac{n\pi y}{y_1} \right\} \left\{ \sum_{n=1}^{\infty} \frac{1}{n} e^{-[n\pi/z_1]^2 \alpha t} \sin \frac{n\pi}{z_1} \right\}$$

and the boundary conditions on the surface of the box are the same as those for the one-dimensional solutions for the infinite plate.

References

1. J. Fourier, *Théorie de la propagation de la chaleur*, 1807, printed, with comments, in *Joseph Fourier 1768–1830*, by I. Grattan-Guinness, MIT Press, 1972.
2. Kaye and Laby, *Tables of physical and chemical constants*. Longman, 1986.
3. Y. S. Touloukian *et al.*, *Thermal Conductivity*, Vols. 1, 2 and 3. IFI/Plenum, 1970.
4. D. Tabor, *Gases, liquids and solids and other states of matter*. Cambridge University Press, 1991.
5. W. M. Rohsenow, J. P. Hartnett and E. N. Ganic, *Handbook of heat transfer fundamentals*. McGraw-Hill, 1985.
6. H. S. Carslaw and J. C. Jaeger, *Conduction of heat in solids*. Oxford University Press, 1959.

Problems

A number of these problems also involve the use of the convective heat transfer coefficient h; where needed the value of h is given, though it has, in fact, been calculated using the methods of the next two chapters. For this reason some of the information given in the questions might appear to be irrelevant.

2.1. A composite wall consists of a 110 mm layer of brick, thermal conductivity 0.7 W m^{-1} K^{-1}, followed by a 20 mm layer of plaster, conductivity 0.47 W m^{-1} K^{-1}. What thickness of insulation, $k = 0.04$ W m^{-1} K^{-1}, should be added to bring the heat loss through the wall down to only 10% of the previous value? Ignore any convective resistance to the heat flow to the air on either side of the wall.   *[72 mm]*

2.2. A horizontal steel pipe of bore 50 mm, with walls 5 mm thick, contains water at 80 °C, flowing at 1 m s^{-1}. The surroundings are still air at 20 °C.

Assuming that the thermal conductivity of the steel is 50 W m^{-1} K^{-1} and the internal and external heat transfer coefficients are 5500 W m^{-2} K^{-1} and 5.5 W m^{-2} K^{-1} respectively, how much heat is lost per metre length of pipe and what are the internal and external wall temperatures?

[62.1 W, 79.9 °C, 79.9°C]

What is the main factor in determining the heat loss?

2.3. The steel pipe of question 2.2 is now covered with a layer of insulation 25 mm thick, of thermal conductivity 0.05 W m^{-1} K^{-1}. Repeat the calculation of heat loss. Assume the same inside coefficient but an outside heat transfer coefficient of 3.3 W m^{-2} K^{-1}. [21.4 W]

What now is the main factor in determining the heat loss?

2.4. A certain nuclear reactor uses solid cylindrical uranium dioxide fuel. The diameter of the fuel region is 13.5 mm. The length of the fuel rods is very long compared to their radius so it is reasonable to assume purely radial heat flow. At a certain position the surface temperature of the fuel is 600 °C and the heat generation rate is 140 Wcm^{-3}. Find the maximum temperature in the fuel at this position, given that the thermal conductivity of UO$_2$ is 2.8 W m^{-1} K^{-1}.

[1170 °C]

If the maximum permissible temperature in the fuel is 1700 °C what is the highest heat generation rate that can be used? (270 W cm^{-3})

2.5. A metal conductor, of rectangular cross-section 12.5 mm by 150 mm, carries an electric current of 10 000 amp. The outer surfaces of the conductor are at 150 °C. Find the value and position of the maximum temperature inside the conductor. Assume one-dimensional heat conduction, i.e. ignore heat transfer through the short sides. The thermal conductivity of the metal is 12 W m^{-1} K^{-1} and its electrical resistivity ρ is 5 $\times$ 10^{-7} Ω m. (electrical resistance is $\rho L/A$). [173.1 °C]

2.6. Show that the steady state one-dimensional differential heat conduction equation may be satisfied by expressions of the following forms

a) $T = a$ b) $T = a + bx$ c) $T = a + bx + cx^2$

In each case state whether the solution corresponds to heat generation or no heat generation and suggest what real problem might give rise to these solutions. a, b and c are constants.

d) Repeat the question but for the transient one-dimensional equation without heat generation and the expression $T = ae^{-bx}\sin(\omega t - bx) + c$. How is b related to the diffusivity?

2.7. In order to measure the thermal diffusivity of soil two temperature sensors are buried in the ground at depths of 50 and 100 mm. In the daily records of temperature it is found that the upper sensor shows a maximum temperature at 3.05 p.m. and the lower at 4.09. Find the diffusivity. Assume that the daily fluctuations of temperature are sinusoidal. [1.17 10^{-6} m^2 s^{-1}]

2.8. A transformer, together with its casing, may be modelled as a vertical cylinder, 0.3 m diameter by 0.4 m high. It generates 400 W of heat. Cooling is effectively only from the vertical sides by natural convection of the air with a heat transfer coefficient of 5.6 W m^{-2} K^{-1}. The ambient air temperature is 20 °C.

a) What is the temperature of the surface of the casing?   [209 °C]

To improve the cooling longitudinal aluminium fins, 1 mm thick and 10 mm long, with a 6 mm pitch, are added to the outside of the casing. Assuming *h* is unchanged,

b) What is the fin efficiency (thermal conductivity of aluminium is 200 W m^{-1} K^{-1})?   [99.8%]

c) What is the new casing surface temperature?   [64 °C]

2.9. With the same basic geometry and heat output as in question 7, would fins be worthwhile if the casing and fins were made of stainless steel (thermal conductivity 12 W m^{-1} K^{-1}) and the casing immersed in a circulating oil bath giving a heat transfer coefficient of 8000 W m^{-2} K^{-1}?

a) assuming the same fin geometry as in question 7?

[No, fin efficiency is only 8.2%]

b) assuming any reasonable geometry?   [No]

2.10. A stage in the manufacture of a certain metal product involves heating the assembly up in a furnace to cause the various parts to be brazed together. The product can be approximated by a cube of 2 m side with an effective thermal conductivity of 10 W m^{-1} K^{-1} and density of 500 kg m^{-3}. Specific heat capacity is 900 J kg^{-1} K^{-1}. Damage to the product will occur if the temperature anywhere exceeds 320 °C.

If the assembly starts at a temperature of 20 °C and its surface is instantaneously raised in the furnace to 300 °C how long will be take to bring the centre temperature up to 290 °C? Hint: assume that only the first term in the infinite series is important but check the relative size of the second term.

[6.85 hours, 2nd term < 10^{-5} of 1st]

3 Forced convection

3.1 Introduction

Convection is the term used to describe the heat transfer process when there is a flow of fluid taking the heat from one place to another. This flow may arise from buoyancy forces, as in natural convection, the subject of the next chapter, or it may be caused by a pump or fan, in other words forced convection, the subject of this chapter. It is worth distinguishing between convection as a mechanism locally within the flow and the overall heat transfer process, which is also called convection.

Locally within a turbulent flow energy is moved from one point to another by movement of the turbulent eddies. Since quite large regions of fluid can be involved in these movements and the turbulent velocities are significant, this method of heat transfer can be much more effective than simply moving energy from one molecule to the next (conduction). The overall heat transfer process is also called convection, even if this turbulent mixing is absent or unimportant.

When turbulent mixing is significant, for normal coolants, the main barrier to heat transfer lies in a thin sub-layer of slow-moving fluid next to the solid surface. The flow in this layer is laminar and heat transfer across it is by conduction. So we have the slightly anomalous situation that the mechanism of conduction is always important in convection heat transfer. The term convection is still used though, one reason, being, that in turbulent flows, although we know that conduction across the laminar layer by the solid wall is important, this information is not particularly helpful if we don't know how thick the layer is. Also the flow of fresh fluid past the surface is essential otherwise the heat transfer would very soon cease.

Since we have already started talking about laminar and turbulent flows it would be as well to revise some basic ideas.

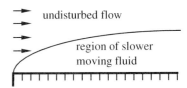

Fig. 3.1 Development of boundary layer on a flat plate

Background fluid mechanics

A particularly important concept is that of the **boundary layer**, the region of the flow close to the solid surface where the flow has been disturbed by the presence of the solid surface. Fig. 3.1 shows the development of the boundary layer on a flat plate. The flow is parallel to the plate surface. The velocity in contact with the solid surface is always zero.

In **laminar flow** the fluid flows in layers which slide past one another and do not mix (Fig. 3.2). Steady streamlines exist (and can be made visible using traces of smoke or dye). Laminar flow is more likely to occur with slow moving, viscous, fluids. In the context of the boundary layer on the plate, Fig. 3.1, laminar flow occurs at the start of the boundary layer, near the leading edge, and also very close to the wall (even if the rest of the boundary layer has become turbulent.)

In **turbulent flow**, superimposed on the main forward movement, are turbulent eddies which tend to mix the fluid (Fig. 3.3). There are no longer any

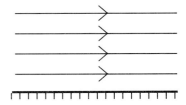

Fig. 3.2 Laminar flow

Fig. 3.3 Turbulent flow

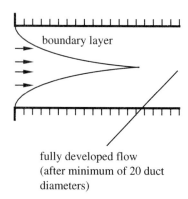

fully developed flow
(after minimum of 20 duct
diameters)

Fig. 3.4   Flow development in a duct

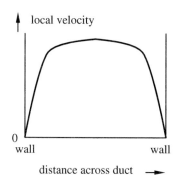

Fig. 3.5   Velocity profile for fully
developed turbulent flow in a duct

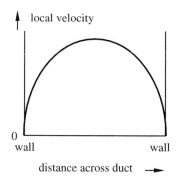

Fig. 3.6   Velocity profile for fully
developed laminar flow in a duct

distinct streamlines; two particles passing the same point will follow different paths. Any dye or smoke introduced to make the streamlines visible will be mixed with surrounding fluid and tend to disappear.

For flow in a duct (Fig. 3.4), initially, near the inlet, a boundary layer similar to that on a flat plate is observed. Eventually the boundary layers meet in the centre of the duct. This is fully developed flow; no further change occurs. The main flow may be laminar or turbulent, depending on the value of a dimensionless group, the Reynolds number

$$Re = \rho u d / \mu \tag{3.1}$$

where ρ is density of the fluid, u the average velocity, d a characteristic linear dimension of the flow geometry, and μ the viscosity. For flow in a circular tube d is just the diameter and the transition from laminar (low Re) to turbulent flow occurs at a Reynolds number of around 3000. Even when the main flow is turbulent there is a laminar sub-layer very close to the wall.

In turbulent flow, because of the mixing, the velocity is fairly constant throughout most of the flow cross-section. Close to the wall, in the laminar sub-layer, it falls rapidly to zero. A typical velocity profile for developed flow in a duct (observed after a length equal to about 20 duct diameters) is shown in Fig. 3.5. These velocities are, of course, averaged over time. The instantaneous values will fluctuate.

In laminar flow the effect of viscosity is to give a steady variation of velocity, as in Fig. 3.6 (again for developed flow in a duct, though the entrance length required is much greater than for turbulent flow).

Heat transfer in laminar forced convection
Since there is no mixing, heat can only be transferred by conduction. The calculation is, however, more complicated than it would be for a solid. Not only is there the question of the geometry to consider but varying velocities affect the amount of heat going to different regions of the flow (more heat will be carried away by a high velocity region). It is the shape of the velocity profile which matters; the heat transfer does not change with the value of the mean velocity, since the average distance the heat has to be conducted is unchanged.

Heat transfer in turbulent convection—importance of velocity
In many practical situations the main flow is turbulent. Within the turbulent core the mixing gives good heat transfer. The main barrier to heat transfer is conduction through the laminar sub-layer, since gases and most liquids are poor thermal conductors (exception: liquid metals).

Heat transfer is mainly determined by the thickness of the laminar sub-layer. A higher flow velocity means a reduced sub-layer thickness, i.e. improved heat transfer. In other words, a layer of warm fluid tends to build up near the heat transfer surface impeding further heat transfer. A higher fluid velocity sweeps this layer away.

3.2   Heat transfer coefficient

If the temperature at the outer surface of the solid is T_w and the bulk or mixed mean temperature of the coolant is T_b then the **heat transfer coefficient** h is defined by

$$\text{heat flux } = q = h(T_w - T_b) \tag{3.2}$$

(as in the previous chapter).

The heat flux is the rate at which heat flows across unit area of the heat-transfer surface. This definition is just a statement of **Newton's law of cooling**, which is closely obeyed in forced convection.

Newton did not write the law in this form[1] but he was the first to assume that the cooling rate is proportional to the temperature difference. He had only a peripheral interest in establishing a law of convective cooling. His main concern at that time was to establish a temperature scale that extended to high temperatures. His thermometer only went up to a little over 200 °C. To extend the range he used a large, red-hot, iron bar. The bar had small samples of different metals on it that solidified at different temperatures. In the range where he could use his thermometer he showed that the cooling of the bar in a draught of air did obey his law. He could then use the law and the cooling times to deduce the higher temperatures (see problem 3.8).

The use of the heat transfer coefficient in the modern manner started with Fourier (reference 1 in the previous chapter).

The meaning of the coolant **bulk temperature, T_b,** is easy enough to understand in the case of flow of a large volume of coolant around the outside of a comparatively small object. It is simply the undisturbed, or free stream, temperature of the coolant a long way from the object, sometimes written T_∞. In the case of fully developed flow inside a duct, where undisturbed coolant does not exist, the meaning of T_b is slightly more complicated. As Fig.3.7 shows, the coolant flowing past a given point of the heat transfer surface is not all at the same temperature; for a turbulent flow, the coolant temperature falls from T_w in contact with the surface to below T_b in the centre of the channel. The temperature distribution in Fig. 3.7 is closely analogous to the velocity distribution for turbulent flow shown earlier in Fig. 3.5. In each case there is a steep variation in the laminar sub-layer near the wall and the turbulent mixing ensures relatively little change in the core of the flow.

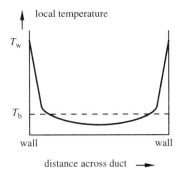

Fig. 3.7 Temperature profile for fully developed turbulent flow in a duct

The bulk temperature T_b may be worked out by considering the temperature of the coolant at the inlet to the duct T_1 and the total heat added up to the point of interest.

$$T_b = T_1 + \text{(rate of heat addition)}/mc \tag{3.3}$$

where m is the mass flow rate and c the specific heat of the coolant at constant pressure. This equation assumes a constant value of c over the temperature range. A more accurate equation is

$$H_b = H_1 + \text{(rate of heat addition)}/m$$

where H stands for the enthalpy of the coolant at the appropriate temperature.

It might seem that in convection the heat transfer coefficient has taken over the role of the thermal conductivity that appears in conduction. This is largely true but there are a couple of differences: k is a physical property of the coolant; h is not—it depends on the velocity as well. Also the use of the thermal conductivity and the Fourier law implies a physical mechanism—conduction. Use of the heat transfer coefficient does not imply a mechanism. There are cases in convection where the mechanism whereby the heat gets into the fluid is entirely conduction. The process is none the less described as convection and a heat transfer coefficient may well be calculated.

Dimensionless groups controlling forced convection heat transfer

Mathematical analysis can be used to find the heat transfer coefficient for laminar flow in simple geometries. (In the next chapter we give an example; even for a simple case the mathematics becomes quite involved.) Since analysis from first principles does not work for all cases, particularly for turbulent flows, it is useful to have an alternative approach. This alternative, dimensional analysis, does not completely solve the problem but it does put some order into the correlation of experimental data.

Even if the heat transfer coefficient cannot be worked out from first principles, any equation for it must satisfy the condition that all the terms in it have the same units and dimensions, and this can be used to find the dimensionless groups of variables that must appear in the equation. However, before the process of dimensional analysis can be started, it is necessary to know all the variables that influence the phenomenon under consideration.

Suppose, as is in fact found by experiment, that the value of the heat transfer coefficient h is determined by the following variables: the coolant viscosity μ, density ρ, thermal conductivity k, specific heat capacity (at constant pressure) c, the flow velocity u, and some characteristic dimension of the flow channel d. The geometry of the flow channel will also influence the heat-transfer coefficient, so the argument is confined to one particular geometry which can be completely defined by a single linear dimension d.

These parameters we have met before, apart perhaps from viscosity. This is related to the shear stress in a (laminar) fluid flow and is defined by shear stress equals viscosity times velocity gradient.

Whether or not h turns out to be expressible as a simple analytical function of these variables it must be possible to write it as an infinite series

$$h = B_1 \mu^\alpha \rho^\beta k^\gamma c^\delta u^\varepsilon d^\zeta + B_2 \mu^{\alpha'} \rho^{\beta'} k^{\gamma'} c^{\delta'} u^{\varepsilon'} d^{\zeta'} + \text{similar terms} \qquad (3.4)$$

where B_1, B_2,etc. are dimensionless constants, and each term of the series must have the dimensions of heat transfer coefficient. The dimensions of the variables, in terms of mass M, length L, time t, and temperature T, are:

$$\begin{array}{ll} h & Mt^{-3}T^{-1} \qquad c \quad L^2t^{-2}T^{-1} \\ \mu & ML^{-1}t^{-1} \qquad u \quad Lt^{-1} \\ \rho & ML^{-3} \qquad\quad d \quad L \\ k & MLt^{-3}T^{-1} \end{array}$$

(These dimensions can be deduced from the units of the various quantities, fairly obviously in the case of d and u, less simply for the quantities involving energy; units of energy are J = N m = kg m s^{-2} m or kg m^2 s^{-2}. We could conduct the argument in terms of the SI units of the various quantities but since the conclusion is valid in any consistent set of units it is better to use M, L, t and T.)

Considering the first term in the infinite series for h (eqn. 3.4) and applying the condition that it must have the same dimensions as h to each of the fundamental quantities in turn:

$$\text{equating dimensions of } M \quad 1 = \alpha + \beta + \gamma \tag{3.5}$$

i.e. h, μ, ρ and k all have dimension M to the power 1, and they appear in equation 3.4 to the powers 1, α, β, and γ respectively. The other variables do not depend on M.

$$\text{Equating dimensions of } L \qquad 0 = -\alpha - 3\beta + \gamma + 2\delta + \varepsilon + \zeta \tag{3.6}$$

$$\text{Equating dimensions of } t \qquad -3 = -\alpha - 3\gamma - 2\delta - \varepsilon \tag{3.7}$$

$$\text{Equating dimensions of } T \qquad -1 = -\gamma - \delta \tag{3.8}$$

With four equations and six unknown powers clearly the information is insufficient to solve the problem completely. However the four equations can be used to reduce the number of unknowns by four. Arbitrarily choosing α and δ as the quantities to keep, and expressing all the others in terms of these two:

$$\text{equation 3.8 gives} \qquad \gamma = 1 - \delta \tag{3.9}$$

$$\text{equation 3.5 gives} \qquad \beta = 1 - \alpha - \gamma = -\alpha + \delta \tag{3.10}$$

$$\text{equation 3.7 gives} \qquad \varepsilon = -\alpha + \delta \text{ (after simplification)} \tag{3.11}$$

$$\text{equation 3.6 gives} \qquad \zeta = -1 - \alpha + \delta \text{ (after simplification)} \tag{3.12}$$

and substituting these expressions back in eqn. 3.4 gives

$$h = B_1 \mu^\alpha \rho^{-\alpha+\delta} k^{1-\delta} c^\delta u^{-\alpha+\delta} d^{-1-\alpha+\delta} + \text{ similar terms}$$

This equation can be rearranged as

$$\frac{hd}{k} = B_1 \left(\frac{\rho u d}{\mu} \right)^{-\alpha+\delta} \left(\frac{\mu c}{k} \right)^\delta + \text{ similar terms} \tag{3.13}$$

Now hd/k is a dimensionless group called the Nusselt number, Nu. $\rho u d/\mu$ is the Reynolds number, Re. $\mu c/k$ is the Prandtl number, Pr. So it is possible to express the heat transfer by

$$Nu = f(Re, Pr) \tag{3.14}$$

The process of dimensional analysis reveals nothing about the form of the function f since in principle there may be an infinite number of terms in equation 3.13 and each one may have different values of α and δ.

With practice in performing this kind of analysis it becomes clear that the number of dimensionless groups formed is normally equal to the number of the original variables minus the number of fundamental quantities that are used, and also that by solving the equations in a different order it is possible to arrive at different dimensionless groups as solutions to the same problem. For example, if the Nusselt number is divided by $RePr$ a new dimensionless group, the Stanton number ($St = h/\rho uc$)is formed, and equation 3.14 shows that

$$St = Nu/RePr = Re^{-1}Pr^{-1}f(Re,Pr) = f'(Re,Pr). \tag{3.15}$$

Experimental measurements on heat transfer are normally presented in terms of the dimensionless groups, since this greatly reduces the volume of data. Over a limited range of Reynolds and Prandtl numbers it is normally possible to correlate the results with adequate accuracy by a simple equation. This equation can then be used to predict the heat-transfer coefficient for another situation, which might involve a different coolant or a different size of channel, provided the Reynolds and Prandtl numbers are within the range for which the correlation has been tested.

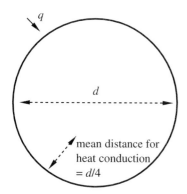

Fig. 3.8 Heat conduction in laminar flow in a duct (approximate)

Physical significance of the dimensionless groups

Since each of the groups is dimensionless it is possible to think of it as the ratio of similar quantities.

Consider first the Nusselt number. Conduction is important either in a laminar flow (conduction across much of the flow cross-section) or in a turbulent flow (conduction across the laminar sub-layer).

In laminar flow we can make a reasonable guess of the average distance the heat has to be conducted and assume that at this position the local fluid temperature is T_b. For the specific case of a circular tube, diameter d, as shown in Fig. 3.8, we could say that the heat has to be conducted, on average, halfway to the centre line, i.e. a distance of $d/4$. The Fourier law gives the heat flux, which in turn equals $h\Delta T$, as roughly

$$q = k(T_w - T_b)/(d/4) = h(T_w - T_b)$$

so $k/h = d/4$ and we can interpret the Nusselt number as a ratio of lengths

$$Nu = \frac{hd}{k} = \frac{d}{k/h} = \frac{\text{characteristic dimension}}{\text{average distance for heat conduction}} = \frac{d}{d/4} = 4 \quad (3.16)$$

Somewhat fortuitously this is close to accurately calculated values for laminar flow in a circular tube (4.36 for constant heat flux at the wall; 3.66 for constant temperature). Later we give accurate values for other geometries.

For turbulent flow the Nusselt number is related to the laminar sub-layer thickness, δ. Very roughly, approximating the real, smooth variation of temperature in a turbulent flow by a core with a constant temperature T_b and a linear change of temperature over the sub-layer then (Fig. 3.9), the Fourier law would give the heat flux

$$q = k(T_w - T_b)/\delta$$

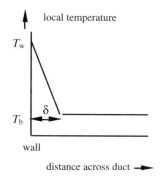

Fig. 3.9 Heat conduction in turbulent flow in a duct (approximate)

and using the definition of h $\qquad q = h(T_w - T_b)$

So again $\quad Nu = \dfrac{hd}{k} = \dfrac{\text{characteristic dimension}}{\text{average distance for heat conduction}} = \dfrac{d}{\delta} \quad (3.17)$

The Reynolds number controls the nature of the flow in the channel, i.e. the degree of turbulence and the flow pattern. It is the ratio of the inertial forces tending to push the flow forwards, ρu^2 per unit area, to the viscous forces tending to slow it down, which are proportional to viscosity times velocity gradient, or roughly to $\mu u/d$ per unit area (this follows from the definition of molecular viscosity).

The Prandtl number determines the effect of the fluid properties on the heat transfer. It is the ratio of the molecular diffusivities of momentum and heat. In transient heat-conduction problems the quantity $k/\rho c$ appears, the thermal diffusivity α (as mentioned in the previous chapter). It has the units $m^2 \, s^{-1}$ and determines how far the heat can diffuse in a certain time. Equally the quantity μ/ρ, which also has the units $m^2 \, s^{-1}$, controls the diffusion of momentum.

3.3   Heat transfer correlations for external flow (i.e. flow around the outside of objects)

Flow past a flat plate
For laminar flow in the boundary layer a theoretical analysis is possible. The mathematics are quite complicated. Details are given, for example, by Schlichting[2]. The average heat transfer, based on the plate length L in the flow direction and the free stream velocity, is given by

$$Nu = 0.664 Re^{1/2} Pr^{1/3} \qquad (3.18)$$

i.e. $Nu = hL/k$ and $Re = \rho u_\infty L/\mu$. The equation is effectively valid for $0.6 < Pr < \infty$. Transition to a turbulent boundary layer occurs at around $Re = 5 \times 10^5$.

It is recommended that the properties are calculated at the **film temperature** T_f which is the average of the wall and bulk temperatures, i.e. $T_f = (T_w + T_b)/2$.

Equation 3.18 is not valid for liquid metals. For $Pr \ll 1$ the analysis gives

$$Nu = 1.13(RePr)^{1/2} \qquad (3.19)$$

Returning to ordinary fluids, for turbulent flow reference 3 recommends

$$Nu = 0.036 Pr^{0.43}(Re^{0.8} - 9200)(\mu/\mu_w)^{1/4} \qquad (3.20)$$

where properties, apart from μ_w, are worked out at the free stream (i.e. bulk) temperature. The 9200 term allows for the laminar behaviour on the early part of the plate. The equation is valid for $0.7 < Pr < 380$ and the upper limit on Re is 5×10^6.

Example 3.1
Oil at 40 °C flows at 0.8 m s^{-1} over a 1 m long plate. The surface temperature of the plate is 80 °C. Find the average heat transfer coefficient and the total heat transferred to the oil if the plate is 0.7 m wide. Properties of the oil at 60 °C: density 880 kg m^{-3}, viscosity 0.20 kg m^{-1}s^{-1}, specific heat 1900 J kg^{-1} K^{-1} and thermal conductivity 0.14 Wm^{-1} K^{-1}.

Taking the properties as given we find $Re = \rho Lu/\mu = 880 \times 1 \times 0.8/0.2 = 3520$ and $Pr = \mu c/k = 0.2 \times 1900/0.14 = 2714$.

The Reynolds number is much less than 5×10^5 so the boundary layer flow is laminar. With such a large Prandtl number equation 3.18 applies. We note that the properties for this equation should be at the film temperature $T_f = (T_w + T_b)/2 = (80 + 40)/2 = 60$ °C, so the properties used are correct. Using equation 3.18

$$Nu = 0.664 \, Re^{1/2} Pr^{1/3} = 0.664 \times 3520^{1/2} \times 2714^{1/3} = 550$$

and the heat transfer coefficient is

$$h = Nu \, k/L = 550 \times 0.14/1 = 76.9 \qquad \text{W m}^{-2} \text{ K}^{-1}$$

The total rate of heat loss is

heat flux $\times$ area $= h(T_w - T_b) \times$ area $= 76.9 \times (80-40) \times 1 \times 0.7 = 2150$ W

Flow past a sphere

The flow around bluff objects such as spheres or cylinders is complex. Only at very low Reynolds numbers does the main flow stay in contact with the surface all the way round. At higher velocities the main flow separates. This gives large local variations in heat transfer.

The results that follow are averages for the entire surface. An equation given by Whitaker[3] is

$$Nu = 2 + \{0.4Re^{1/2} + 0.06Re^{2/3}\}Pr^{0.4}(\mu/\mu_w)^{1/4} \qquad (3.21)$$

where Nu and Re are based on the sphere diameter, and all properties, apart from μ_w, are taken at the free stream temperature, T_∞. If the wall temperature is unknown then the equation should first be used without the correction term, μ/μ_w, to make an estimate of T_w.

The first term in the equation, i.e. $Nu = 2$, is the steady state conduction solution found in the previous chapter. The range of the data used was $4 < Re < 7 \times 10^4$ and $0.71 < Pr < 380$.

For liquid metals ($Pr \ll 1$) a suggested equation [4], based on analysis and in reasonable agreement with experimental data, is

$$Nu = 2 + 0.775(RePr)^{0.5} \qquad (3.22)$$

Flow normal to a cylinder

An equation that has been proposed[3] is very similar to that for a sphere

$$Nu = \{0.4Re^{1/2} + 0.06Re^{2/3}\}Pr^{0.4}(\mu/\mu_w)^{1/4} \qquad (3.23)$$

The range of the data used was $40 < Re < 10^5$ and $0.67 < Pr < 300$. Again the characteristic length is the diameter. At very small Re values the equation tends to $Nu = 0$. This is the correct theoretical result for conduction from an infinitely long cylinder but measurements usually find that Nu tends to a small finite value. Taking the limiting value mentioned in Reference 3 we can modify the equation to

$$Nu = 0.24 + \{0.4Re^{1/2} + 0.06Re^{2/3}\}Pr^{0.4}(\mu/\mu_w)^{1/4} \qquad (3.24)$$

Noting in addition that this equation is in reasonable agreement with that recommended in 5, that is valid at higher Re values, we can suggest equation 3.24 to cover $0.02 < Re < 10^6$.

For liquid metals a suggested[6] correlation is

$$Nu = 1.125(RePr)^{0.413} \qquad (3.25)$$

for $1 \leq RePr \leq 100$.

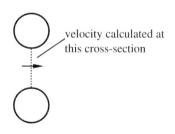

Fig. 3.10 Reynolds number is calculated with the maximum velocity

velocity calculated at this cross-section

Flow over tube banks

We assume a non-metallic fluid and that the flow is normal to the cylindrical tubes, i.e. crossflow. The heat transfer for the first row of tubes is similar to that of isolated tubes. For the later tubes the turbulence generated by the first row tends to give enhanced heat transfer. This is particularly noticeable at high Reynolds numbers.

It is normal to define the Reynolds number using the maximum velocity in the tube bank (Fig. 3.10), that is the velocity through the narrowest cross-section. A thorough study has been made by Zukauskas.[5]. We note first of all that the single cylinder correlation recommended in reference 5 gives similar results to equation 3.24. In what follows we have considerably simplified the recommendations of Zukauskas but at no point have we approximated by more than 10%.

We consider first tube banks with the tubes in-line (see Fig. 3.11). The expressions which follow are for the average heat transfer in a tube bundle consisting of many rows (strictly 20 or more).

At low Reynolds numbers, between 10 and 1000, the single tube equation 3.24 can be used. This does not mean that the heat transfer is the same as for an isolated tube because the Reynolds number is now formed with the maximum velocity. This has the effect of increasing the Nusselt number.

At higher Reynolds numbers this procedure underestimates the heat transfer so we take the following equations from reference 5:

$$10^3 < Re < 2 \times 10^5 \qquad Nu = 0.27Re^{0.63}Pr^{0.36}(Pr/Pr_w)^{0.25} \qquad (3.26)$$

$$2 \times 10^5 < Re < 10^6 \qquad Nu = 0.021Re^{0.84}Pr^{0.36}(Pr/Pr_w)^{0.25} \qquad (3.27)$$

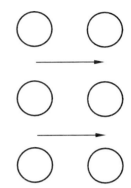

Fig. 3.11 Heat exchanger with tubes in line

Again properties, apart from Pr_w, are calculated at the bulk temperature, Nu and Re are formed with the tube diameter, and the maximum velocity is used to find Re.

For tube banks of less than 20 rows the average heat transfer is reduced, but not very much. Even for 4 rows, depending on Reynolds number, the reduction is a maximum of 10%. For bundles with 3, 2 or a single row of tubes, with $Re > 10^3$, the reduction is about 14, 20 or 31% respectively.

For staggered tube banks, as in Fig. 3.12, the heat transfer is on average about 10% higher. So the Nu number calculated from the previous results for the in-line tube bundle should be increased by a factor of 1.1. This suggests that staggered tube banks should always be preferred. However the pressure drop across staggered banks is also higher so the decision is less clear cut.

3.4 Heat transfer correlations for turbulent flow of non-metals in ducts

Circular tubes
For fully developed turbulent flow of a gas or non-metallic liquid in a smooth tube, diameter d, the most widely quoted equation is that of Dittus and Boelter, and reference 7 is given, which is odd because the equation does not appear in the reference! Nonetheless the following has become known as the Dittus–Boelter equation:

$$Nu = 0.023Re^{0.8}Pr^{0.4} \qquad (3.28)$$

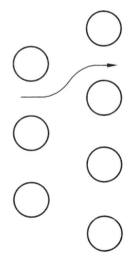

Fig. 3.12 Heat exchanger with staggered tubes

The fluid properties are calculated at the bulk coolant temperature. In this context 'fully developed' means that the length of duct is more than 20 times the diameter and 'turbulent flow' means $Re = \rho du/\mu > 10^4$. In spite of its simplicity the equation gives good results for gases and, provided the Prandtl and Reynolds numbers are not too high, for liquids also. The equation should not be used if there are large variations of temperature and hence properties between the bulk coolant and the heated surface. More detailed discussion of all of these points is given later.

A comparison of the equation with the experimental results of a dozen different groups of workers, over a wide range of Reynolds numbers, and Prandtl numbers up to 100, showed an overall root-mean-square error between the individual experimental points and the predictions of the equation of about 10%.[8] More accurate and more generally applicable equations exist [9,10] but even these can differ from one another and from experimental results by 10%.

Incidentally a slightly different version of the Dittus–Boelter equation is sometimes recommended specifically for heating of the tube wall, i.e. the fluid temperature is higher than the wall temperature. In this alternative version $Pr^{0.4}$ is replaced by $Pr^{0.3}$. This alternative version does not appear to be more accurate. The recommended equations for the case of large differences between the wall and bulk temperatures are given later.

Entrance effects

The correlation of experimental results, equation 3.28, is for fully developed turbulent flow. It applies therefore to the flow some distance from the inlet to the·channel where the boundary layer has built up to its equilibrium thickness. At the entrance to the channel, or following a sharp bend or sudden reduction in cross-section, the boundary layer is likely to be thinner and the heat transfer coefficient as much as two or three times the equilibrium value. This high value only persists for one or two channel diameters downstream.

Non-isothermal flow

Because of the importance of the laminar sublayer in controlling the heat transfer rate some other correlations prefer to evaluate the properties at the film temperature i.e. $T_f = (T_w + T_b)/2$. This can have the disadvantage that the wall temperature is not known in advance, so a value of T_w will have to be guessed and iteration used. To cope more accurately with all these effects the following equations are suggested.

The Gnielinski equations

These equations have been fitted to experimental results[10] Apart from the correction factor for non-isothermal flow they are again based on properties evaluated at the bulk fluid temperature. They are valid for $2300 < Re < 10^6$. d is the channel diameter and L is the channel length. The equations give an average heat transfer coefficient over the length L.

For $0.6 < Pr < 1.5$ (usually gases) :

$$Nu = 0.0214(Re^{0.8} - 100)Pr^{0.4}\left[1 + \left(\frac{d}{L}\right)^{2/3}\right]\left(\frac{T}{T_w}\right)^{0.45} \tag{3.29a}$$

T_b is the bulk temperature and T_w the wall temperature (in K).

For $1.5 < Pr < 500$ (usually liquids):

$$Nu = 0.012(Re^{0.87} - 280)Pr^{0.4}\left[1 + \left(\frac{d}{L}\right)^{2/3}\right]\left(\frac{Pr}{Pr_w}\right)^{0.11} \tag{3.29b}$$

Pr is the bulk value and Pr_w the value at the wall temperature. If the wall temperature is not known in advance then iteration must be used.

Equivalent diameter

A simple approximate method for other shapes of channel is to use the equivalent diameter defined by

$$d_e = \frac{4 \times \text{ flow cross-sectional area}}{\text{perimeter of channel}} \tag{3.30}$$

For flow inside a circular tube we have a flow area of $\pi d^2/4$ and a perimeter of πd so this definition gives the actual diameter.

For a number of simple geometries the equivalent diameter can be used in equation 3.28 (or 3.29) to give heat-transfer coefficients within 10 or 20%. The reason for the success of this procedure is not entirely clear. Certainly a recent numerical study[11] predicted virtually identical Nusselt numbers for flow in circular tubes and between flat plates, based on the equivalent diameter. One might suppose that the procedure would also work for flow in any duct shape in between, such as squares, rectangles or annuli. This is confirmed experimentally[12]. For triangular ducts, though, the method overestimates the heat transfer.

Comparison with the equations in the next section shows that the Dittus–Boelter equation, with d_e, can still be used to give a rough estimate even for flow outside tube bundles but tends to underestimate the value of Nu at the widest lattice spacings.

This suggests the limitations of the Dittus–Boelter equation: it tends to overestimate the heat transfer in confined geometries (e.g. inside a triangular duct) and it tends to underestimate it in very open geometries (e.g. outside widely spaced tubes).

Correlation for water flow parallel to the outside of tube bundles

A correlation for this specific case has been derived by Weisman[13]; it has been confirmed by more recent data[14].

$$Nu = C \, Re^{0.8} Pr^{1/3} \tag{3.31}$$

For a triangular array $C = 0.026P/D - 0.006$ valid for $1.1 \leq P/D \leq 1.5$

For a square array $C = 0.042P/D - 0.024$ valid for $1.1 \leq P/D \leq 1.3$

where P/D is the pitch to diameter ratio. (Note that even though the correlation has been specifically developed for a particular channel shape the Reynolds number and Nusselt number are still worked out using the equivalent diameter, d_e.)

Example 3.2

Air at an average pressure of 10 bar and a bulk temperature of 300 °C flows through a 0.5 m long, 20 mm diameter, circular tube at 2.7 m^3 hr^{-1}. The heat flux on the tube wall is 12 kW m^{-2}. What is the heat transfer coefficient and what is the inside wall temperature?

Either equation 3.28 or equation 3.29 can be used. There is much to be said for using both, as a check and because the more accurate equation, 3.29, is difficult to use until we have at least an estimate of the wall temperature.

The flow rate information is given as a volume flow rate, i.e. 2.7 / 3600 = $uA = u\pi r^2 = u\pi \, 0.01^2$ giving velocity $u = 2.387$ m s^{-1}.

Taking values from the appendix the Reynolds number is

$$Re = \frac{\rho d u}{\mu} = \frac{6.077 \times 0.02 \times 2.387}{2.938 \times 10^{-5}} = 9875$$

The Prandtl number from the table is 0.693, so substituting in the Dittus–Boelter equation

$$Nu = 0.023 Re^{0.8} Pr^{0.4} = 31.16$$

and the heat transfer coefficient $h = Nu\ k/d = 31.16 \times 0.04417\ /\ 0.02 = 68.8$ W m^{-2} K^{-1}. The inside wall temperature is $300 + q/h = 300 + 1.2 \times 10^4\ /\ 68.8 = 474\ °C$

We can now check these results and perhaps get more accurate values using the Gnielinski equation:

$$Nu = 0.0214(Re^{0.8} - 100)Pr^{0.4}\left[1 + \left(\frac{d}{L}\right)^{2/3}\right]\left(\frac{T}{T_w}\right)^{0.45}$$

$$= 0.0214(9875^{0.8} - 100)0.693^{0.4}\left[1 + \left(\frac{0.02}{0.5}\right)^{2/3}\right]\left(\frac{573}{747}\right)^{0.45} = 26.91$$

Repeating the previous calculations gives $h = 59.4$ W m^{-2} K^{-1}, a change of 14% from the previous value, and $T_w = 502\ °C$.

3.5   Correlations for turbulent flow of liquid metals in ducts

Constant heat flux to slug flow in a circular tube
Although this model is approximate it gives a usable result for the Nusselt number and it illustrates, in a simple way, the type of calculation that is done when the heat transfer mechanism is purely conduction.

Since in turbulent flow the velocity is nearly constant over the most of the flow cross-section it is a reasonable approximation to say that the velocity is constant everywhere. In other words the liquid behaves like a solid slug of metal as it moves through the tube.

At some distance from the inlet to the channel the temperature profile in the metal slug or rod will have settled down to an equilibrium shape. The temperature is still rising in the rod but at every radius it is rising at the same rate, and the centre-to-edge temperature difference stays constant. With the temperature rising at the same rate at all radial positions the heat supplied to each region of the liquid metal must be in proportion to the volume of metal. If the heat flux is q then the total rate of heat supply per unit length of the channel is $2\pi aq$, and the heat flowing into the region of the liquid metal bounded by the cylindrical surface radius r (Fig. 3.13), volume πr^2, is

$$2\pi aq\ \pi r^2/(\pi a^2)$$

where a is the radius of the tube. This heat must be conducted through the surface at r, so from the Fourier law

$$2\pi aqr^2/a^2 = k2\pi r\ dT/dr$$

(the minus sign in the Fourier law has been cancelled since the expression for the rate of heat flow represents heat flowing inwards, i.e. in the negative r direction).

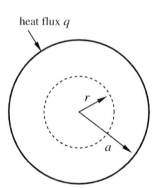

heat flux q

Fig. 3.13   Slug flow model. Heat flow through surface at r is proportional to volume inside

To find the bulk temperature, T_b, we first have to integrate to get the temperature

$$\frac{q}{ka} \int_r^a r\,dr = \int_{T_r}^{T_w} dt$$

i.e.

$$\frac{q}{ka}[(a^2 - r^2)/2] = T_w - T_r$$

The bulk temperature of the flow is just the volumetric average temperature, since the velocity is the same everywhere:

$$T_b = \frac{1}{\pi a^2} \int_0^a T_r 2\pi r\,dr = T_w - \frac{q}{ka^3} \int_0^a (a^2 - r^2)r\,dr$$

or

$$T_w - T_b = \frac{q}{ka^3}\left[\frac{a^4}{2} - \frac{a^4}{4}\right] = \frac{qa}{4k}$$

The heat transfer coefficient h is $q/(T_w - T_b) = 4k/a$, so the Nusselt number is

$$Nu = \frac{hd}{k} = \frac{4k}{a}\frac{2a}{k} = 8 \tag{3.32}$$

Accurate analysis for turbulent flow in a circular tube
An equation for constant heat flux proposed by Lyon,[15] following on from a similar earlier analysis by Martinelli[16] is

$$Nu = 7 + 0.025(RePr)^{0.8} \tag{3.33}$$

Both analyses were numerical; the equation is fitted to the numerical results. Fully developed flow is assumed and both conduction (the constant 7 term) and convection (the term with $RePr$) taken into account. The reason for the conduction contribution giving $Nu = 7$ rather than 8 is that the actual turbulent velocity profile was used instead of assuming a constant velocity. The combination $RePr$ which frequently appears in liquid metal heat transfer may be replaced by the Peclet number ($Pe = RePr$).

For a highly conducting liquid metal such as sodium the conduction term can dominate even at quite high velocities (hence the claim that $Nu = 8$ is a usable result!).

For liquid metals the boundary condition on the wall assumes greater importance than for non metals, controlling as it docs the temperature distribution throughout the flow. For a constant temperature boundary condition the 7 term in equation 3.33 should be replaced by 5[17].

It should be recognised that measurements of liquid metal heat transfer have often yielded Nusselt numbers well below those of the Lyon–Martinelli equation (3.33), sometimes as low as half the theoretical value, and empirical correlations are often recommended that give lower values. There is evidence that these lower values may be due to a small degree of contamination of the heat transfer surface, masking the inherently good heat transfer of the liquid metal. Measurements under carefully controlled conditions[18], with provision for measuring temperature profiles within the liquid metal flow as well as in the tube wall, so that any contact resistance can be detected, show good agreement with equation 3.33.

Quantitative explanations that have been suggested for the surface contamination include deposits of oxides[19] and layers of gas bubbles on the surface[20]. These effects are more likely at low temperatures and when the liquid metal does not wet the surface well.

Flow parallel to rod bundles
An empirical equation for triangular lattices with constant heat flux is[21]

$$Nu = 0.25 + 6.2P/D + (0.032P/D - 0.007)(RePr)^{(0.8-0.024P/D)} \qquad (3.34)$$

valid for $150 < RePr < 3000$ and $1.2 < P/D < 2$. P/D is the pitch to diameter ratio and Nu and Re are based on the equivalent diameter.

3.6   Laminar flow in ducts

For fully developed laminar flow in ducts, since there is no turbulent mixing, analysis of the heat conduction process is possible. A very simple example of the type of reasoning involved was given in the previous section. Although we were talking there about turbulent flow the slug flow model assumed that the effect of turbulent mixing was negligible. The same slug flow assumption would be a poor one for laminar flow because the velocity is far from constant (e.g. Fig. 3.6). It is necessary to include the varying velocity in the calculations. Since the velocity is higher in the centre of the duct more heat has to flow to the centre (compared to slug flow), the average distance the heat has to be conducted is higher and the Nusselt number is lower.

In Fig. 3.14 the Nusselt number values for fully developed flow for various geometries are given.[22] These are for the boundary condition of uniform heat flux in the flow direction and uniform wall temperature at any axial position. The Nusselt number is based on the equivalent diameter. In problem 3.5 we argue that the equivalent diameter is not a very logical choice for a purely conduction problem, but it is the convention to use it.

Nusselt numbers for uniform wall temperature are about 20% lower than those given in Fig. 3.14. For a circular tube $Nu = 3.66$.

A problem with fully developed laminar flow is that it often doesn't exist. Entrance lengths are much longer than for turbulent flow. Near the entrance the heat transfer coefficient is higher because the heat only has to be conducted a short distance. A simple equation for the average heat transfer in the entrance region of a circular tube[23] is

$$Nu = 1.86Re^{1/3}Pr^{1/3}(L/D)^{-1/3}(\mu/\mu_{\rm w})^{0.14} \qquad (3.35)$$

with properties, apart from $\mu_{\rm w}$, at the bulk temperature. At sufficiently long tube length L this equation will predict Nu below the fully developed value. At this point it can be assumed that the flow is fully developed and the Nu values in Fig. 3.14 apply.

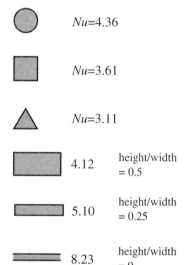

$Nu=4.36$

$Nu=3.61$

$Nu=3.11$

4.12   height/width = 0.5

5.10   height/width = 0.25

8.23   height/width = 0

Fig. 3.14   Nusselt number for fully developed laminar flow in various geometries

3.7   Temperature transients—the lumped parameter approximation

As mentioned in Chapter 2 the analysis of the temperatures in a solid object during a transient can be complex. One limiting case of the behaviour is seen when the heat conduction within the object is very efficient compared to the convection of heat from the surface. This might well apply to a metal object that is cooled, or heated, at its surface by a gas. The temperature within the object is essentially uniform and the rate at which temperature changes with time is determined by the convective heat transfer at the surface. The temperature of the solid is treated as a lumped parameter, i.e. for the purposes of the analysis the solid becomes a single lump with a single temperature.

We assume that the heat transfer coefficient, h, and the physical properties are constant. This is likely to be valid in forced convection if the range of temperature is not too great and the absolute temperature does not become so high that radiation has to be taken into account. The shape of the solid region is irrelevant (except as it will affect the calculation of h). The cooling case is considered where the object starts at time $t = 0$ at a high temperature T_0 and is suddenly placed in a different environment where it is cooled by surrounding fluid at a constant temperature T_b. Obviously the temperature, T, of the solid will eventually reach T_b. This type of analysis was introduced by Newton[1].

As shown in Fig. 3.15 the volume of the solid is V and its surface area A. The rate of heat loss from the surface is heat flux times area or

$$h(T - T_b)A$$

(since the temperature of the wall is just T).

The heat lost in time dt is $h(T - T_b)A dt$ and this equals the change in stored heat as the temperature falls by dT, i.e.

$$V\rho c\, dT = -hA(T - T_b)dt \tag{3.36}$$

defining $\theta = T - T_b$ then $d\theta = dT$ and

$$\frac{d\theta}{\theta} = -\frac{hA\, dt}{V\rho c}$$

Integrating between time zero and an arbitrary time t gives

$$[ln\theta]_{\theta_0}^{\theta} = -\frac{hA}{V\rho c}[t]_0^t$$

i.e.

$$\frac{\theta}{\theta_0} = e^{-\frac{hAt}{V\rho c}}$$

or

$$\frac{T - T_b}{T_0 - T_b} = e^{-\frac{hAt}{V\rho c}} \tag{3.37}$$

As expected the temperature T eventually falls to T_b. The time constant for the process is $V\rho c/hA$, i.e. the temperature difference falls to $1/e$ of its initial value in this time.

Although we stated that we were considering the cooling case, the mathematics are quite general and the same equation, 3.37, applies for heating. If $T_b > T$ then $T - T_b$ is a negative quantity and equation 3.36 is still correct, showing that the temperature is now increasing with time. Equation 3.37 still results, although both numerator and denominator take negative values.

Validity of lumped approach—Biot number

The lumped parameter approach will be valid provided the temperature differences within the solid are much less than those outside.

Within the solid there is a heat flux due to conduction which can be written

$$q = k\, dT/dx \approx k\Delta T_{solid}/L \tag{3.38}$$

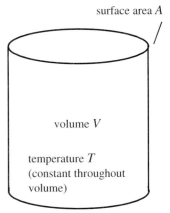

surface area A

volume V

temperature T
(constant throughout volume)

Fig. 3.15 Variables for the lumped parameter analysis. Analysis applies to any shape

where L is a characteristic linear dimension.

From the surface there is a heat flux due to convection

$$q = h(T - T_b) \tag{3.39}$$

and these two estimates of heat flux are roughly equal (it is the heat flow due to conduction which supplies the heat flow due to convection at the surface).

We require $\Delta T_{\text{solid}} \ll (T - T_b)$

and replacing the temperatures using equations 3.38 and 3.39 this becomes

$$qL/k \ll q/h$$

or

$$hL/k \ll 1$$

Now

$$hL/k = Bi = \textbf{the Biot number} \tag{3.40}$$

and the condition for the lumped parameter approach to be valid is **Bi << 1**. We can regard the dimensionless group of the Biot number as the ratio of resistance to heat conduction inside the solid to resistance to heat convection from the surface. Note that k here is the value for the solid (and this is the difference between the Biot number and Nusselt number definitions).

Example 3.3

A thermocouple is being used to measure the temperature of a stream of air. The temperature of the air suddenly changes from 20 °C to 100 °C. If the thermocouple junction can be modelled as a 3 mm diameter sphere, with negligible conduction down the wires, how long will it take for the registered temperature to reach a) 90 °C, b) 99 °C?

The velocity of the gas stream is 0.9 m s^{-1}. Metal properties are: density 8600 kg m^{-3}, specific heat 400 J kg^{-1} K^{-1} and thermal conductivity 30 W m^{-1} K^{-1}. Show that the lumped parameter approach is justified.

We anticipate that equation 3.37 will apply but we need a value of the heat transfer coefficient. The bulk air temperature is 100 °C throughout so we take air properties at 100 °C in order to be able to use the equation for the heat transfer from a sphere, equation 3.21.

We have $Re = \rho Du/\mu = 0.9456 \times 0.003 \times 0.9/2.183 \times 10^{-5} = 117.0$

and the Prandtl number for air at 100° C is 0.702 so we are virtually within the valid range of equation 3.21. Ignoring the viscosity correction for the moment we have

$$Nu = 2 + \{0.4 \times 117^{1/2} + 0.06 \times 117^{2/3}\}0.702^{0.4} = 7.002$$

To use the viscosity correction we need to know the wall temperature. Unfortunately this varies during the temperature transient, and a varying heat transfer coefficient is not consistent with the lumped parameter analysis. The best that can be done is to calculate a reasonable average value of h. The mean of the initial and final values of T_w is 60° C so

$$(\mu/\mu_w^{1/4}) = (2.183 \times 10^{-5}/2.008 \times 10^{-5})^{1/4} = 1.021$$

The corrected value of Nu is 7.002 $\times 1.021 = 7.107$.

$$h = Nu \, k/D = 7.107 \times 0.031 \, 31/0.003 = 74.17 \qquad \text{W m}^{-2} \text{ K}^{-1}.$$

The error in assuming a constant h is small: the change in h during the heating process is only a couple of per cent.

The parameter that comes into equation 3.37 is

$$hA/V\rho c = h4\pi r^2/(\rho c 4\pi r^3/3) = 3h/(\rho cr) = 3 \times 74.17/(8600 \times 400 \times 0.0015)$$
$$= 0.0431 \text{ s}^{-1}$$

a) The time t to reach 90 °C is given by equation 3.37. Taking the natural logarithm of each side the equation becomes

$$-(hA/V\rho c)t = \ln\{(T - T_b)/(T_0 - T_b)\}$$

so $\qquad\qquad -0.0431t = \ln\{(90 - 100) - (20/100)\} = \ln\{10/80\}$

and $\qquad\qquad\quad t = 48.2 \text{ s}.$

b) Repeating the calculation with $T = 99$ °C gives a time of 101.7 s.

To check that the lumped parameter approach is valid the Biot number is found:

$$Bi = hD/k = 74.2 \times 0.003/30 = 0.0074 \ll 1$$

3.8 Heat exchangers—Log mean temperature difference

A very simple heat exchanger could be just a double pipe as shown in Fig. 3.16. This is the simplest representation of a counter flow heat exchanger. To avoid using lots of subscripts let the bulk temperature of the hot fluid in the central pipe be T and the bulk temperature of the cooler fluid flowing in the outer annulus be θ (these are not the same definitions used in the previous section). The purpose of the heat exchanger, of course, is to use the hotter fluid to warm the cooler fluid. The driving force for heat exchange at any position is the temperature difference $T - \theta = \Delta T$.

With full information on fluid flow rates and properties the heat transfer coefficient on each side of the pipe wall could be calculated, plus any conduction resistance through the wall itself. We assume that this has been done and an overall heat transfer coefficient U (as in Chapter 2) found, such that the local heat flow rate is

$$dQ = UdA(T - \theta) = UdA\Delta T \qquad (3.41)$$

where A is the total heat transfer area in the heat exchanger and dA an element of the area. We assume that U is constant.

For the case where $T - \theta$ also stays constant throughout the heat exchanger (perhaps from exact matching of the flow rates) then the total rate of heat transfer is simply $Q = UA(T - \theta) = UA\Delta T$.

The problem is that normally the driving temperature difference ΔT changes. We suppose that all the inlet and the outlet temperatures are known, i.e. $\Delta T_1 = T_1 - \theta_1$ and $\Delta T_2 = T_2 - \theta_2$ are known. It is tempting to say that it is sufficient to use the average temperature difference $\Delta T_{av} = (\Delta T_1 + \Delta T_2)/2$ and thus

$$Q = UA\Delta T_{av} \qquad (3.42)$$

This gives reasonable results provided ΔT_1 and ΔT_2 are within a factor of two of one another but it is not quite correct.

The correct starting point is equation 3.41. To integrate it we need information on how ΔT varies along the heat exchanger. It is helpful to be able to us an expression like 3.42 but we need the correct expression for the average temperature difference. To anticipate the result this is called the log mean temperature difference, i.e.

$$Q = UA\Delta T_{\mathrm{LMTD}} \tag{3.43}$$

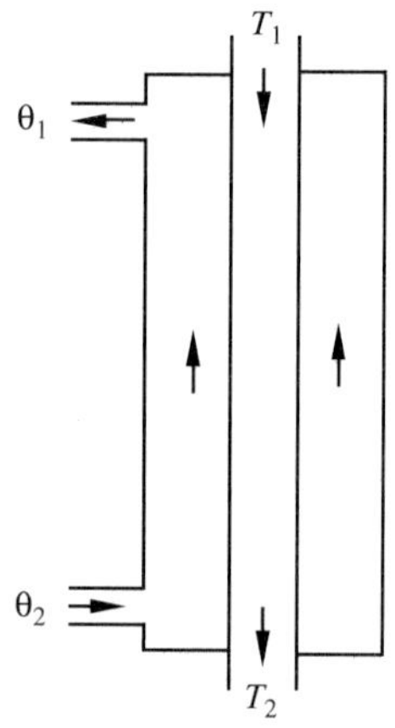

Locally the heat exchanged, equation 3.41, has the effect of changing the temperature of the fluids. Working from position 1 to position 2, i.e. from top to bottom of the heat exchanger as drawn (Fig. 3.16), on each side the fluid temperature is falling

$$dQ = UdA\Delta T = -m_c c_c d\theta \tag{3.44}$$

and

$$dQ = UdA\Delta T = -m_h c_h dT \tag{3.45}$$

where m_c and c_c are mass flow rate and specific heat of the cooler fluid, and similarly for the hotter fluid.

Also equations 3.44 and 3.45 can be integrated along the channel to give

$$Q = m_c c_c (\theta_1 - \theta_2) \tag{3.46}$$

and

$$Q = m_c c_c (T_1 - T_2) \tag{3.47}$$

To return to trying to solve equation 3.41, information on $\Delta T = T - \theta$ comes from rearranging equations 3.44 and 3.45 and subtracting, i.e. $d\theta = UdA\Delta T / m_c c_c = -dT = UdA\Delta T / m_h c_h$ giving

$$d(\Delta T) = dT - d\theta = UdA\Delta T \left\{ \frac{1}{m_c c_c} - \frac{1}{m_h c_h} \right\}$$

and the *mc* terms can be replaced using equations 3.46 and 3.47, giving

$$d(\Delta T) = \frac{UdA\Delta T}{Q} \{\theta_1 - \theta_2 - T_1 + T_2\} = \frac{UdA\Delta T}{Q} \{-\Delta T_1 + \Delta T_2\}$$

$$\text{or} \quad \int_1^2 \frac{d(\Delta T)}{\Delta T} = \int_1^2 \frac{U}{Q} \{-\Delta T_1 + \Delta T_2\} dA$$

$$\text{i.e.} \quad \ln{[\Delta T]_1^2} = \ln\left[\frac{\Delta T_2}{\Delta T_1} \right] = \frac{U}{Q} \{-\Delta T_1 + \Delta T_2\} A$$

and comparing this with equation 3.43 we see that the appropriate temperature difference is the log mean temperature difference given by

$$\Delta T_{\mathrm{LMTD}} = \frac{\{\Delta T_2 - \Delta T_1\}}{\ln\left[\frac{\Delta T_2}{\Delta T_1} \right]} \tag{3.48}$$

This is a rather odd looking expression and it is not immediately easy to check it; it does not obviously reduce to $\Delta T_{\mathrm{LMTD}} = \Delta T_1 = \Delta T_2$ for the limiting case of $\Delta T_1 = \Delta T_2$. Those of you who enjoy mathematical manipulation and can remember the expansion $\ln{(1 + x)} = x$ for small x can try substituting $\Delta T_2 = \Delta T_1 (1 + x)$ in equation 3.48 to find the limit.

It is possible to show that equation 3.48 is also valid for co-current flow, i.e. the two fluids in Fig. 3.16 both going in the same direction.

Fig. 3.16   Simple counterflow heat exchanger

A practical heat exchanger will almost certainly require more surface area than is provided by the double tube arrangement of Fig. 3.16. One possibility is several tubes inside a shell, Fig. 3.17, i.e. a shell and tube heat exchanger. One fluid flows through the tubes, the other around the outside of the tubes. To the extent that the fluid flow paths are parallel equation 3.48 still applies.

Just to emphasise that many types of heat exchanger exist, Fig. 3.18 shows a plate-fin heat exchanger. The different fluids are separated by plates. The heat transfer area is increased by corrugations between the plates, brazed to the plates. The corrugations constitute fins, which extend each of the plate surfaces. Many more layers would be present than shown in the figure. One possibility, as shown, is to have the flows in adjacent layers at right angles.

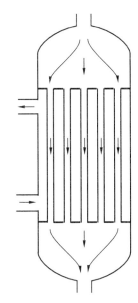

Fig. 3.17 Shell and tube heat exchanger. The two flows are separate, one inside the tubes, one around the outside

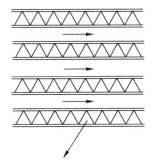

Fig. 3.18 Plate fin heat exchanger. The corrugations, and flows, are at right angles in adjacent layers

References

1. I. Newton, *Table of quantities and degrees of heat,* 1701, reprinted and translated in J. F. Scott, *The correspondence of Isaac Newton.* Cambridge University Press, 1967.
2. H. Schlichting, *Boundary-layer theory* (7th edn). McGraw-Hill, 1979 (reissued 1987).
3. S. Whitaker, Forced convection heat transfer correlations for flow in pipes, past flat plates, single cylinders, singles spheres and for flow in packed beds and tube bundles. *A. I. Ch. . Journal* **18**, 361–371, 1972.
4. G. Refai Ahmed and M. M. Yovanovich, Approximate analytical solution of forced convection heat transfer from an isothermal sphere for all Prandtl numbers. *ASME J. Heat Transfer* **116**, 838–843 1994.
5. A. Zukauskas, Heat transfer from tubes in crossflow. *Adv. Heat Transfer* **8**, 93–160, 1972.
6. R. Ishiguro, K. Sugiyama, and T. Kumada, Heat transfer around a circular cylinder in a liquid-sodium crossflow, *Int. J. Heat Mass Transfer* **22**, 1041–1048, 1979.
7. F. W. Dittus and L. M. K. Boelter, Heat transfer in automobile radiators of the tubular type. *Universtiy of California, Berkeley, Publ. Eng.,* **2**, 443–461, 1930.
8. *Engineering Sciences Data, Heat Transfer, vol.1*, item 67016, 1967.
9. B. S. Petukhov, Heat transfer and friction in turbulent pipe flow with variable physical properties. *Adv. Heat Transfer,* **6**, 504–564, 1970.
10. V. Gnielinski, New equations for heat and mass transfer in turbulent pipe and channel flow, *Int. Chem. Eng.* **16**, 359–368, 1976.
11. N. Ghariban, A. Haji-Sheikh, and S. M. You, Pressure drop and heat transfer in turbulent duct flow: a 2 parameter variational method. *ASME J. Heat Transfer,* **117**, 289–295, 1995.
12. References to several early reports are given in S. T. Hsu, *Engineering Heat Transfer,* van Nostrand, 1963.
13. J. Weisman, Heat transfer to water flowing parallel to tube bundles. *Nucl. Sci. Engng.* **6**, 78–79, 1959.
14. S.-H. Kim and M. El-Genk, Heat transfer experiments for low flow of water in rod bundles. *Int. J. Heat Mass Transfer* **32**, 1321–1336, 1989.
15. R. N. Lyon, Liquid metal heat transfer coefficients. *Chem. Eng. Progress* **47**, 75, 1951.

16. R. C. Martinelli, Heat transmission to molten metals. *Trans. ASME* **69**, 947–959, 1947.

17. R. A. Seban and T. Shimazaki, Heat transfer to fluid flowing turbulently in a smooth pipe with walls at constant temperature, *Trans. ASME* **73**, 803, 1951.

18. P. L. Kirillov *et al.*, Heat transfer in pipes to sodium-potassium alloy and to mercury. *J. Nucl. Energy Part B, Reactor Technology*, **1**, 123–129, 1959.

19. V. I. Subbotin, F. A. Kozlov, and N. N. Ivanovskii, Heat transfer to sodium under conditions of free and forced convection and when oxides are deposited on the surface, High Temperature, **1**, 368-372, 1963.

20. R. H. S. Winterton, Effect of gas bubbles on liquid metal heat transfer. *Int. J. Heat Mass Transfer*, **17**, 549–554, 1974.

21. H. Graber and M. Rieber, Experimental study of liquid metal heat transfer, flowing in line through tube bundles, Progress in Heat and Mass Transfer **7**, 151–166, 1973.

22. R. K. Shah and A. L. London, *Laminar flow: forced convection in ducts*, Academic Press, New York, 1978.

23. E. N Sieder and G. E. Tate, Heat transfer and pressure drop of liquids in tubes. *Ind. Eng. Chem.*, **28**, 1429, 1936.

Problems

3.1. Water at 280 °C flows through a 15 mm i.d. circular tube at 5 m s^{-1}. Calculate the heat transfer coefficient (the pressure is sufficient to prevent boiling).

   If heat is supplied to the tube at a rate of 20 kW per metre length, what is the temperature of the inside surface of the tube?

[34 400 W m^{-2} K^{-1}, 292.3 °C]

   Compare with the prediction of the Gnielinski equation for the average h over a 1 m entrance length (average bulk temperature is still 280 °C).

[33 600 W m^{-2} K^{-1}, 2.5% change]

3.2. A car radiator is required to dissipate 9 kW when travelling at 80 km/hr with ambient air temperatures up to 35 °C. The radiator is made of metal tubes, 7 mm outside diameter, 0.5 mm wall and 500 mm long, arranged so that the air flow is perpendicular to the tubes. Fins are added to the outside of the tubes. The effect of the fins is to increase the effective area of the outside of the tubes by a factor of 5. If pumping limitations suggest a water flow velocity of 0.9 m s^{-1} and the water temperature is 60 °C how many tubes are needed?

[Around 37 tubes]

   What assumption has to be made?

3.3. Air at 5 bar and 200 °C flows at 10 m/s along a tube of internal diameter 25 mm. Locally the wall temperature is 30 K above the air temperature. Calculate the heat transfer per unit length of tube at this position. If this heat transfer rate were maintained all along the length of the tube how much would the bulk temperature increase over a 2 m length?   [311.5 W m^{-1}, 33.8 K]

3.4. A 5 μm diameter tungsten wire is being used to measure the velocity of an atmospheric pressure air flow. A 2 mm length is supported perpendicluar to the flow. Associated electronics are used to keep the resistance, i.e. temperature, of the wire constant at 50 °C. If the air temperature and velocity are 20 °C and 40 ms^{-1} respectively what voltage will be registered? Will voltage be proportional to velocity?   [0.231 volts; No]

Resistivity ρ of wire is 6.1 $\times$ 10^{-8} ohm m. Resistance is given by $R = \rho L/A$.

3.5. In Fig. 3.14 the Nusselt numbers are given for developed laminar flow for a certain boundary condition. Without attempting any detailed analysis make a reasonable estimate of the Nusselt number for the circular and rectangular ducts. Assume that the average distance the heat has to travel is half of the distance from the longer side to the centre and that this is the position where the temperature in the coolant equals the bulk temperature. Hence show that the Nusselt number (based on the shorter side, or on the diameter) is 4.

Next calculate the true Nusselt numbers (based on equivalent diameter) and find the errors compared to the values in Fig. 3.14.
[circle –8%, square 11%, rectangles, starting nearest to square; 29, 25 and –3%]

3.6. A thermocouple is being used to measure the temperature of a stream of air. The temperature of the gas suddenly changes from 20 °C to 200 °C. If the thermocouple junction can be modelled as a 2 mm diameter sphere, with negligible conduction down the wires, how long will it take for the registered temperature to reach a) 100 °C, b) 199 °C?

The gas velocity is 2 m s^{-1}. The metal properties are: density 8900 kg m^{-3}, specific heat 410 J kg^{-1} K^{-1} and thermal conductivity 23 W m^{-1} K^{-1}. Show that the lumped parameter approach is justified.   [a 9.4 s, b 16 s]

3.7. A 10 cm diameter copper sphere at a uniform temperature of 300 °C is suddenly quenched by immersion in a bath of liquid at 40 °C. The heat transfer coefficient is 200 W m^{-2} K^{-1}. Check that it is reasonable to regard the copper temperature as uniform at all times and calculate the time for the sphere temperature to reach 60 °C.

Copper properties are: density 8954 kg m^{-3}, specific heat 383 Jkg^{-1} K^{-1} and thermal conductivity 386 W m^{-1} K^{-1}.   [12.2 min.]

3.8. Professor Newton wishes to measure the melting points of two small samples of metal, A and B. Unfortunately he does not have access to his laboratory and the thermometer he has found only measures up to 100 °C.

He decides to heat up a large iron bar until it is red hot (but still solid). He places it in a steady draught of air at 10 °C with the small samples of metal (now molten) resting on top. He observes the following:

Time (min)	0	4	24	48
Event/Temp.	A solidifies	B solidifies	100 °C	35 °C

What are the temperatures at which A and B solidify? What assumptions are needed to obtain a solution?   [334 and 272 °C]

4 Natural convection

4.1 Introduction

In natural convection the fluid velocity arises purely from the heating or cooling of the heat transfer surface itself. Heating creates less dense fluid next to the surface which tends to rise under buoyancy. The low velocities (compared to forced convection) give low heat transfer coefficients.

Everyday applications of natural convection include space heating from a radiator, the coil dissipating heat to the room from a refrigerator and the cooling of many electronic components.

The hot fluid next to the heated surface experiences a net upward force due to the difference in density between the hot fluid and the surrounding cold fluid. The flow pattern in a typical heated layer is shown in Fig. 4.1. The horizontal scale has been exaggerated by about 5 times. Although it is conventional to show turbulent eddies in this sort of way it should be appreciated that they are superimposed upon a basically upward flow.

Analysis when the flow is turbulent is difficult. There has been some success with simple analysis when the flow is laminar. We will devote some time to analysing laminar flow next to a vertical heated plate. Not only is this interesting in its own right, as one of the relatively few simple examples of producing a usable convective heat transfer relation from first principles, but it gives insight into quantities that are not mentioned in the empirical heat transfer correlations. For example, the thickness of the boundary layer next to the plate is relevant when we are estimating how close fins can be placed on a heat transfer surface. The recommended equations for the vertical plate are given at the end of the analysis.

A confusing factor in the literature is the large number of correlations that have been proposed. In this chapter we have tried to keep to just one correlation for each situation but it should be recognized that other correlations exist in the literature that could well yield heat transfer coefficients 5 or 10% different (as mentioned in the first chapter). Also the correlations tend to apply to idealized geometries. For example, the equations might strictly apply to an infinitely long cylinder surrounded by an indefinite body of stationary fluid at a uniform temperature. If the equation is being used to make a prediction for a cylinder of finite length, with other objects close by, there is little point in worrying about which of two or three different correlations might be the best to use.

In the heat transfer correlations that are presented later in this chapter it is assumed that the physical properties that are required are calculated at the film temperature.

Fig. 4.1 Air velocity next to a 1 m high heated vertical surface

turbulent

laminar

4.2 Laminar natural convection on a vertical, isothermal plate

For this case a reasonably accurate analytical solution is possible. We assume that the region of disturbed velocity and temperature is confined to a well defined boundary layer and that the thickness of the boundary layer is the same for both thermal and velocity effects. For simplicity we will talk about a heated plate with buoyant fluid rising up next to it, but the final equations are equally valid for a cooled plate.

The main approximation is that we do not attempt to find the detailed distribution of velocity and temperature within the boundary layer; instead we fit simple polynomials to the boundary conditions. Since the polynomials give a smooth variation, without rapid changes near the wall, the solution is appropriate for laminar flow.

As shown in Fig. 4.2 the distance x is measured from the bottom edge of the plate. Consider the balance of forces on a element dx of the boundary layer, between the plate and the edge of the boundary layer, and of width unity.

The edge of the boundary layer merges smoothly with the undisturbed fluid so the velocity at the edge of the boundary layer, at $y = \delta$, is given by $u = du/dy = 0$. On the surface of the plate, as shown in Fig. 4.3, the viscous force opposing the upward flow is, from Newton's law of viscosity

$$\mu \left(\frac{du}{dy} \right)_{y=0} dx \tag{4.1}$$

and at the edge of the boundary layer there is no viscous force since $du/dy = 0$.

The volume of the element is δdx (times unity) where δ is the thickness of the boundary layer at this position. The weight of the fluid in it is $\rho_{av} g \delta dx$ where ρ_{av} is the average density of the hot fluid in the boundary layer. The buoyancy force created by the weight of the displaced fluid is $\rho_c g \delta dx$, where ρ_c is the density of the undisturbed bulk fluid, so the net force is

$$g \delta dx (\rho_c - \rho_{av}) - \mu \left(\frac{du}{dy} \right)_{y=0} dx$$

The density of hot fluid is related to that of the cold fluid by

$$\rho = \rho_c (1 - \beta[T - T_\infty])$$

where β is the volume coefficient of expansion. The average density difference is

$$\rho_c - \rho_{av} = \frac{1}{\delta} \int_0^\delta \rho_c \beta [T - T_\infty] dy = \frac{\beta \rho}{\delta} \int_0^\delta [T - T_\infty] dy \tag{4.2}$$

where, in the last term, by assuming that the differences of temperature are small compared to the absolute level, β and ρ_c can be approximated by the mean values (to be evaluated in due course at the film temperature) and taken outside the integral.

To complete the momentum equation, in the form net force equals rate of change of momentum, we note that the standard expression linking mass flow rate m and velocity u is $m = \rho u \times$ flow cross-sectional area. So the mass flow rate in a strip dy is $\rho u dy$ and the rate at which momentum is being carried is $\rho u^2 dy$, giving a total momentum of $\int \rho u^2 dy$ and the change in this over dx is $d[\int \rho u^2 dy]$. So the momentum equation is

$$g \delta dx \frac{\beta \rho}{\delta} \int_0^\delta [T - T_\infty] dy - \mu \left(\frac{du}{dy} \right)_{y=0} dx = d \int_0^\delta \rho u^2 dy = \rho d \int_0^\delta u^2 dy \tag{4.3}$$

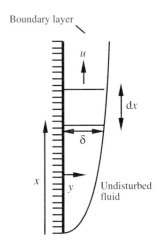

Fig. 4.2 Parameters used in analysis of boundary layer

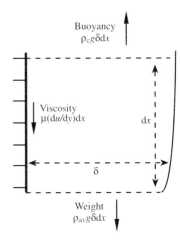

Fig. 4.3 Force balance on a slice dx of boundary layer

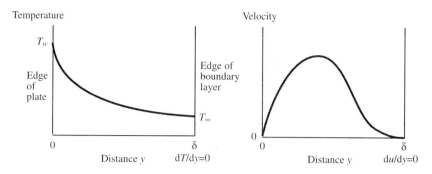

Fig. 4.4   Shape of the temperature and velocity profiles is essentially determined by fit to boundary conditions

The energy equation comes from equating the heat being conducted into the boundary layer at the plate surface to that required to warm up the expanding boundary layer flow. Again we note that the mass flow rate in the strip dy is $\rho u\,dy$ and so the rate at which heat is being carried above the undisturbed temperature is $\rho u c[T - T_\infty]dy$, c being the specific heat capacity.

$$-k\left(\frac{dT}{dy}\right)_{y=0} dx = d\int_0^\delta \rho u c[T - T_\infty]dy = \rho c d\int_0^\delta u[T - T_\infty]dy \qquad (4.4)$$

The fluid beyond $y = \delta$ is unaffected by the heated plate, so there is no heat flow at the edge of the boundary layer, i.e. at $y = \delta$, $T = T_\infty$ and dT/d$y = 0$.

These equations seem complicated but if the equations for the velocity and temperature profiles are known then all the required gradients and integrals can be calculated in terms of T_w, $T\infty$, u_{av} and δ

Looking at Fig. 4.4 it seems reasonable that any equations for the profiles that are forced to meet the boundary conditions will give reasonable representations of the required gradients at $y = 0$ and of the required integrals. The boundary conditions on u and T at the edge of the boundary layer have already been mentioned. In addition, at $y = 0$, $T = T_w$ and $u = 0$.

We represent T by the simplest polynomial that can be expected to fit the variation in the Fig. 4.4

$$T = a_0 + a_1 y + a_2 y^2 \qquad (4.5)$$

$T = T_w$ at $y = 0$ gives $T_w = a_0$
dT/d$y = 0$ at $y = \delta$ gives $0 = a_1 + 2a_2\delta$
$T = T_\infty$ at $y = \delta$ gives $T_\infty = T_w + a_1\delta + a_2\delta^2$

a_0 is known and there are two equations for a_1 and a_2. The temperature profile can finally be expressed as

$$T = T_w - (T_w - T_\infty)\left[2\frac{y}{\delta} - \left(\frac{y}{\delta}\right)^2\right] \qquad (4.6)$$

For the velocity profile, trying the same approach as equation 4.5 does not work; all the coefficients turn out to be zero. One further term in the polynomial is needed:

$$u = a_0' + a_1'y + a_2'y^2 + a_3'y^3 \qquad (4.7)$$

Obviously $a'_0 = 0$ since $u = 0$ at $y = 0$. The remaining two boundary conditions are not sufficient to find the remaining three coefficients. Applying the boundary conditions and rearranging the resulting equations gives

$$\frac{u}{u_0} = \frac{y}{\delta}\left[1 - \frac{y}{\delta}\right]^2 \tag{4.8}$$

where u_0 is related to the remaining coefficient and is a measure of the amplitude of the velocity distribution.

It is now possible to work out all the gradients and integrals required in equations 4.3 and 4.4. The following results are fairly straightforward:

$$u_{av} = \frac{1}{\delta}\int_0^\delta u\,dy = \frac{1}{12}u_0 \qquad \left(\frac{du}{dy}\right)_{y=0} = \frac{u_0}{\delta} = 12\frac{u_{av}}{\delta}$$

$$\left(\frac{dT}{dy}\right)_{y=0} = -\frac{2}{\delta}(T_w - T_\infty) \qquad \frac{1}{\delta}\int_0^\delta[T - T_\infty]dy = \frac{1}{3}(T_w - T_\infty)$$

and the following quite tedious to work out:

$$\int_0^\delta u^2\,dy = \frac{u_0^2\delta}{105} = \frac{144}{105}u_{av}^2\delta \qquad \int_0^\delta u[T - T_\infty]dy = \frac{2}{5}u_{av}[T_w - T_\infty]\delta$$

So equations 4.3 and 4.4 can be simplified to:

$$g\delta dx\beta\rho[T_w - T_\infty] - \mu\frac{36u_{av}}{\delta}dx = \rho d\left(\frac{432}{105}u_{av}^2\delta\right) \tag{4.9}$$

and

$$k\left(\frac{[T_w - T_\infty]}{\delta}\right)dx = \rho c\frac{1}{5}d(u_{av}[T_w - T_\infty]\delta) \tag{4.10}$$

Up to this point the method of solution is the same as that described in many heat transfer textbooks (and due originally to Squire[1]). The next step in the standard solution is to assume that both u and δ are proportional to powers of x. Even with this assumption further analysis is required to establish what the two powers are and what the constants of proportionality are. In what follows a slightly more straightforward approach is adopted that does not require any advance assumptions about the form of the solution.

Solution for large Prandtl number

It is reasonable to suppose that with a sufficiently viscous liquid (large Prandtl number) the acceleration term on the right hand side in the momentum equation (equation 4.9) will be negligible, so the equation can be rearranged to find

$$u_{av} = \delta^2\frac{g\beta\rho[T_w - T_\infty]}{36\mu} \tag{4.11}$$

and substituting this into the energy equation (4.10) gives

$$k\left(\frac{[T_w - T_\infty]}{\delta}\right)dx = \rho c\frac{1}{5}d\left(\delta^3\frac{g\beta\rho[T_w - T_\infty]^2}{36\mu}\right) = \rho c\frac{1}{60}\frac{g\beta\rho[T_w - T_\infty]^2}{\mu}\delta^2d\delta$$

or

$$\delta^3d\delta = \frac{60\mu k}{\rho cg\beta\rho[T_w - T_\infty]}dx$$

Integrating this from the bottom of the plate, $x = 0$, up to some position x gives

$$\int_0^\delta \delta^3 d\delta = \frac{60\mu k}{\rho^2 cg\beta[T_w - T_\infty]} \int_0^x dx$$

hence

$$\delta^4 = \frac{240\mu k}{\rho^2 cg\beta[T_w - T_\infty]} x \qquad (4.12)$$

The local heat flux at position x is given by

$$q = -k\left(\frac{dT}{dy}\right)_{y=0} = \frac{2k}{\delta}(T_w - T_\infty)$$

and since the heat transfer coefficient is defined as $q/\Delta T$ then

$$h = \frac{2k}{\delta}$$

The local Nusselt number is

$$Nu_x = \frac{hx}{k} = 2\left(\frac{\rho^2 cg\beta[T_w - T_\infty]x^3}{240\mu k}\right)^{1/4} \qquad (4.13)$$

or

$$Nu_x = 0.508(Gr_x Pr)^{1/4} \qquad (4.14)$$

where two dimensionless groups have been introduced. $Pr = \mu c/k$, the Prandtl number, appeared in the previous chapter. Gr, the Grashof number, is a new group

$$Gr_x \frac{\rho^2 g\beta[T_w - T_\infty]x^3}{\mu^2} \qquad (4.15)$$

Since the two groups appear as the product in this equation, and in many empirical equations for natural convection heat transfer, we replace them by the Rayleigh number ($Ra = GrPr$).

$$Ra_x = \frac{\rho^2 cg\beta[Tw - T_\infty]x^3}{k\mu} \qquad (4.16)$$

So we can write equation 4.14 as

$$Nu_x = 0.508 Ra_x^{1/4} \qquad (4.17)$$

which is in very good agreement with more accurate, numerical, calculations for $Pr > 100$. This agreement is to some extent fortuitous: the more accurate calculations disagree slightly with the assumed temperature and velocity profiles, also with the assumption that the velocity and temperature boundary layers are of equal thickness.

More generally we might hope to write

$$Nu = CRa^n \qquad (4.18)$$

as an equation that will cover any natural convection situation. Indeed this is frequently done, with the constant C varying slightly with the geometry and the power n being $\frac{1}{4}$ for laminar flow. The equation is extended to turbulent flow with different values of C and n. There is a variation with Prandtl number but, as the next sections will show, it is fairly small for ordinary fluids.

To find the Prandtl number dependence for laminar flow, and to justify our assumption of negligible acceleration of the fluid for large values of the Prandtl number, we next consider the general solution.

General solution

Now that we have a solution for the simplified case of negligible acceleration of the boundary layer, i.e. the right-hand side of equation 4.9 set equal to zero, we could assume that the solution is still valid as a first approximation to the full equations, and use the equations (4.11 and 4.12) for u_{av} and δ to give a first estimate of the acceleration, i.e. the $u_{av}^2 \, \delta$ term in equation 4.9. If we do this we find a rather surprising result: the last term in equation 4.9 is simply a constant times the first term. Further iteration does not change this conclusion.

To short circuit this process we simply accept the result that the last term is a constant factor, say F, times the first term and rewrite the equation

$$g\delta dx \beta \rho[T_w - T_\infty] - \mu \frac{36u_{av}}{\delta} dx = Fg\delta dx \beta \rho[T_w - T_\infty] \qquad (4.19)$$

or $(1 - F)g\delta dx \beta \rho[T_w - T_\infty] - \mu \dfrac{36u_{av}}{\delta} dx = 0$

There is no need to go through all the analysis again. We can use the previous results, simply noting that a $(1\text{-}F)$ factor has now to be associated with the term containing β. So the average velocity is now

$$u_{av} = \delta^2 \frac{g\beta\rho[T_w - T_\infty]}{36\mu}(1 - F) \qquad (4.20)$$

and the boundary layer thickness is given by

$$\delta^4 = \frac{240\mu k}{\rho^2 cg\beta[T_w - T_\infty](1 - F)} x \qquad (4.21)$$

The Nusselt number is

$$Nu_x = \frac{hx}{k} = 2\left(\frac{\rho^2 cg\beta(1 - F)[T_w - T_\infty]x^3}{240\mu k}\right)^{1/4} \qquad (4.22)$$

or in terms of the Rayleigh number

$$Nu_x = 0.508Ra_x^{1/4}(1 - F)^{1/4}$$

To check that the solution is self-consistent, and find the value of F, the two versions of the acceleration term in equations 4.9 and 4.19 are equated, i.e.

$$Fg\delta dx \beta \rho[T_w - T_\infty] = \frac{432}{105}\rho d(u_{av}^2 \delta)$$

Using the equations above to substitute first for u_{av} and then, after the differentiation, for δ gives, after some tidying up.

$$1 - F = \left(1 + \frac{0.952}{Pr}\right)^{-1} \qquad (4.23)$$

So the Nusselt number expression becomes

$$Nu_x = 0.508Ra_x^{1/4}\left(1 + \frac{0.952}{Pr}\right)^{-1/4} \qquad (4.24)$$

and the equations above for velocity and boundary layer thickness can be used with the result for 1-F.

Often there is no requirement to find the local heat transfer coefficient. Just the total heat transfer given by the average heat transfer coefficient is needed. The heat transfer coefficient at distance x from the bottom of the plate is given by equation 4.22, i.e.

$$h = 2k\left(\frac{\rho^2 cg\beta(1-F)[T_\mathrm{w}-T_\infty]}{240\mu k}\right)^{1/4} x^{-1/4} = Cx^{-1/4}$$

The average heat transfer coefficient, because of the $x^{1/4}$ dependence, turns out to be 4/3 times the value at $x = L$,

$$h_\mathrm{av} = \frac{1}{L}\int_0^L Cx^{-1/4}\mathrm{d}x = \frac{4}{3}CL^{-1/4} \tag{4.25}$$

and the average Nusselt number is

$$Nu = \frac{4}{3}0.508Ra^{1/4}\left(1+\frac{0.952}{Pr}\right)^{-1/4}$$

where Nu and Ra are both based on the plate height L. This equation is in agreement with accurate, numerical, solutions of the problem, where the correct velocity and temperature profiles are calculated as part of the solution, to within 10% over a very wide range of Prandtl number. For air it reduces to $Nu = 0.546Ra^{1/4}$ which is about 6% high compared to the accurate solutions.

Recommended equations for vertical plate

Bringing our equation into line with the accurate solution for air ($Pr = 0.70$) the recommended equation for the average heat transfer for air becomes

$$Nu = 0.51Ra^{1/4} \qquad \text{for } 10^4 < Ra < 10^9 \tag{4.26}$$

$$\text{with} \qquad Nu = \frac{hL}{k} \qquad \text{and} \qquad Ra = \frac{\rho^2 cg\beta[T_\mathrm{w}-T_\infty]L^3}{k\mu} \tag{4.27}$$

For other fluids the results can be corrected using

$$\frac{Nu}{Nu_\mathrm{air}} = 1.24\left(1+\frac{0.952}{Pr}\right)^{-1/4} \tag{4.28}$$

These two equations (4.26 and 4.28) give good results over a very wide Pr range, as shown in Fig. 4.5. There are a number of interesting features in the figure. It contains examples of all the different types of natural convection information: predictions based on approximate analysis (the equations we have just derived); results of accurate, numerical, calculations (a separate calculation for each Pr value); experimental data; and a correlation shown to agree with a large body of data. The correlation is more complex than equation 4.26 so we have shown the case of $Ra = 10^4$. At high Ra the correlation essentially coincides with the accurate theory. At low Ra loss of heat by conduction (ignored in the theoretical treatments) has a small influence. Our equations, for fluids such as gases or water, lie in between.

In general the value of the volume expansion coefficient β has to be looked up in tables. However a useful result for gases is that $\beta = 1/T$ where T is the absolute temperature in K.

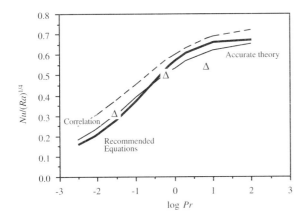

Fig. 4.5 Equations 4.26 and 4.28 compared with numerical results,[3] a correlation (dashed) and experiment[2] (triangles)

The $Ra < 10^9$ condition in equation 4.26 is the condition for laminar flow in the boundary layer. There has been some uncertainty in whether the condition should be on Gr or on Ra, caused by the fact that only a limited range of Pr is normally studied. Recent results, for air[4] and liquid metals [5] on a vertical plate, suggest that the Grashof number is the key parameter for transition from laminar to turbulent conditions. Until this thinking has been extended to other geometries though we have little choice but to take the limits given in the literature, i.e. use the Rayleigh number. At least this has the advantage that we do not need to mention the Grashof number again.

4.3 Turbulent boundary layer

The starting differential equations of the previous section, equations 4.3 and 4.4, still apply so it is reasonable to suppose that the same dimensionless groups will control the heat transfer. Attempts at analysis are made more difficult by the fact that new velocity and temperatures profiles, appropriate to a turbulent flow, must be found.

A simple equation for the overall heat transfer, based on analysis and shown to agree with experimental results for air, was proposed by Bayley[6]:

$$Nu = 0.1\, Ra^{1/3} \qquad \text{for } 10^9 < Ra < 10^{13} \tag{4.29}$$

A recent paper[7] states that other correlations converge around this equation. It appears to give reasonable results for a wide range of fluids (excluding liquid metals). As a result of the role of turbulent mixing in transporting heat and momentum the significance of the Prandtl number is reduced and, for the usual range of fluids at least, no Prandtl number correction is made.

Example 4.1
One side of a 1m high by 2m wide vertical surface at 20 °C is exposed to still air at 10 °C and 1 atm. What is the rate of heat loss due to natural convection?

Firstly we need to find whether conditions are laminar or turbulent, i.e. the value of

$$Ra = \frac{\rho^2 cg\beta[T_\text{w} - T_\infty]L^3}{k\mu} \qquad \text{from equation 4.27}$$

In this geometry the length L is the plate height, 1 m. $T_w = 20$ °C and $T_\infty = 10$ °C.

From the Appendix, interpolating for air at $T_f = 15$ °C, $\rho = 1.226$ kg m^{-3}, $c = 1003$ J kg^{-1} K^{-1}, $k = 0.0252$ W m^{-1} K^{-1}, $\mu = 1.798 \times 10^{-5}$ kg m^{-1} s^{-1}.

In addition we have $g = 9.81$ m s^{-2} and $\beta = 1/(273 + 15) = 0.003\ 472$. Substituting in the Ra expression gives

$$Ra = 1.133 \times 10^9$$

i.e. (just) in the turbulent region so we use equation 4.29 to give

$$Nu = 0.1\ Ra^{1/3} = 104.3$$

and $h = Nuk/L = 2.627$ W m^{-2} K^{-1}

the *rate of heat loss* is heat flux times area or $h(T_w - T_\infty) \times 2 \times 1 = 53\ W$

(note that this last line is the only place that the width of the surface, 2 m, appears; also there will be heat loss due to radiation. In the next chapter we show that this can be calculated independently and added on.)

4.4   Other simple geometries

For the rest of this chapter a couple of simplifications can be made. We will ignore the Prandtl number dependence, i.e. the equations that are quoted are either for air or are based on data that are predominately for air. All the Nusselt number values are averages for the whole surface. As stated at the beginning of the chapter property values are calculated at the film temperature, $T_f = (T_w + T_\infty)/2$. A point to watch is that the definition of the characteristic length in the Nu and Ra numbers (equation 4.27) varies with the geometry.

In laminar flow a correction for different Prandtl numbers can be made using equation 4.28 (there are reasons for thinking that the correction does not depend on geometry: the Prandtl number correction came in right at the beginning of the problem with the original differential equation, 4.3 and 4.9). In turbulent flow any Prandtl number effect appears to be small.

Vertical cylinders of height L give the same results (for their vertical sides) as vertical plates, i.e. equations 4.26 and 4.29 can be used, provided the thickness of the boundary layer is appreciably less than the diameter of the cylinder, D, i.e. using equation 4.21 (with $Pr = 0.7$) the requirement is

$$5Ra^{-1/4} \ll D/L \tag{4.30}$$

For inclined plates one might hope to be able to repeat the analysis of the previous sections with a reduced buoyancy force, replacing g by the component of the acceleration due to gravity parallel to the surface, $g \cos \theta$. θ is the angle between the plate and the vertical, Fig. 4.6. In other words the laminar vertical plate equation is used, with $L =$ the length of the plate (measured in the vertical plane) as before, but with Ra replaced by $Ra \cos\theta$ or

$$Nu = 0.51(Ra\ \cos\theta)^{1/4} \tag{4.31}$$

This works well for downward facing heated plates (where the boundary layer is forced to remain in contact with the plate). The procedure was shown to be valid for laminar flow for θ to within a couple of degrees of the horizontal.[9]

For upward facing heated plates there is a possibility of the boundary layer separating from the plate, the likelihood of which increases as the angle and

Fig. 4.6   Downward facing inclined plate

Ra number increase. It has been shown[10] that the simple procedure of using the vertical plate laminar equation with Ra replaced by $Ra \cos\theta$ works for θ less than 80° and

$$Ra < 2 \times 10^9 \, e^{-0.0513\,\theta}$$

with θ in degrees. For larger Ra the following turbulent equation is recommended ($\theta < 80°$):

$$Nu = (0.1 + \theta/3600)\, Ra^{1/3} \qquad 2 \times 10^9 \, e^{-0.0513\,\theta} < Ra < 10^{10} \qquad (4.32)$$

in which we recognize the Bayley equation (4.29) plus a term giving an increase in heat transfer with angle of inclination. This is the opposite of the laminar equation (4.31) where heat transfer deteriorates with inclination.

For heated horizontal plates facing down (or a cooled plate facing up) a suggested equation[9] is

$$Nu = 0.58 \, Ra^{1/5} \qquad \text{for } 10^6 < Ra < 10^{11} \qquad (4.33)$$

and for heated horizontal plates facing up (or a cooled plate facing down), averaging and slightly approximating two equations from Reference 9

$$Nu = 0.145 \, Ra^{1/3} \qquad \text{for } 10^7 < Ra < 10^{11} \qquad (4.34)$$

Various suggestions have been made for the correct length scale in equations 4.33 and 4.34. The simplest seems to be the side of a square or, equivalently, the shorter side of a rectangle.

An important geometry in practice is the horizontal cylinder. Morgan[11] surveyed the literature and recommended the following equations for long cylinders:

$$Nu = 0.48 \, (Ra)^{1/4} \qquad \text{for } 10^4 < Ra < 10^7 \qquad (4.35)$$

and

$$Nu = 0.125 \, (Ra)^{1/3} \qquad \text{for } 10^7 < Ra < 10^{12} \qquad (4.36)$$

with both Nu and Ra based on cylinder diameter. The data examined by Morgan included other fluids in addition to air. A comparison with more recent data is shown in Fig. 4.7.

Results for a sphere[13], with the diameter as characteristic length, can be expressed as

$$Nu = 2 + 0.43 \, Ra^{1/4} \qquad 1 < Ra < 8 \times 10^8 \qquad (4.37)$$

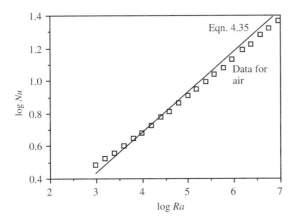

Fig. 4.7 Heat transfer from a horizontal cylinder. Equation 4.35 compared with data from Reference 12

As elsewhere in this section the equation is for air. The water results of Reference 14 are given accurately by this same equation if the recommended Prandtl number correction, equation 4.28, is used (on the second term in 4.37). The upper limit of validity on Ra is based on the water results. The reason for the constant term, 2, in the equation, is discussed later in the section on correlations for lower *Ra* values.

Nearly all the different correlations that have been described for calculating *Nu* show a certain similarity, not to say monotony. Lienhard[15] suggested that, in the laminar range,

$$Nu = 0.52\,Ra^{1/4} \tag{4.38}$$

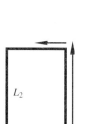

Fig. 4.8 Characteristic length is distance travelled in boundary layer

could be used to estimate the heat transfer for other shapes of bodies for which more specific information was not available. The length scale to be used in calculating *Nu* and *Ra* is the distance a particle travels in the boundary layer, e.g. *L* for a plate, $\pi D/2$ for a cylinder or sphere, $L_1 + L_2$ for the rectangular box shown in Fig. 4.8.

Applied to our previous equations for the vertical plate, cylinder and sphere, equation 4.38 has a maximum error on *Nu* of only 8% (though rather higher on *h*). It does not work very well for inclined or horizontal plates. Work on cubes and squat cylinders suggests that equation 4.38 tends to underestimate the heat transfer[16].

Correlations for lower Ra values

Nearly all of the equations for *Nu* given so far would predict *Nu* = 0 if extrapolated to *Ra* = 0; this is not always correct. The reason is that, provided the body has finite dimensions in all directions, some heat will be lost by conduction to the surroundings even if the natural convection contribution becomes negligible. In the case of conduction from an isothermal sphere into uniform, infinite surroundings, the heat transfer can be expressed as *Nu* = 2 (equation 2.22). This result has already been incorporated in equation 4.37 and explains why it has a much lower limit of validity than all the other equations presented so far.

Probably the most important case where a prediction may be needed is, in practice, for fine wires, i.e. for horizontal cylinders. In reference[17] the following expression is recommended:

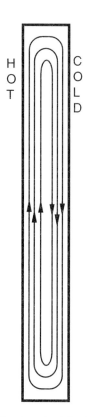

Fig. 4.9 Laminar circulation in cavity

$$Nu = 0.36 + 0.518\left(\frac{Ra}{\left[1 + (0.559/Pr)^{9/16}\right]^{16/9}}\right)^{1/4} \tag{4.39}$$

for $10^{-6} < Ra < 10^{9}$

A slightly unsatisfactory feature of this correlation, recognised in [17], is that the theoretical solution for an infinitely long cylinder has *Nu* = 0 at *Ra* = 0. In practice this is not found, even when end effects are allowed for. A possible explanation is that the apparatus in which the measurements are made is only of finite size.

Enclosed spaces

Another geometry of interest is the enclosed space between two tall vertical plates, e.g. a double-glazed window. A complication here is that although a laminar natural convection circulation may develop (Fig. 4.9), flowing up by the warm surface and down by the cold surface, it does not contribute much to the heat transfer because of the distance the fluid has to travel before it changes direction. Effectively heat transfer is by conduction across the gap over a wide range of conditions. Using the Fourier heat conduction law the heat flux is given by

$$q = k(T_1 - T_2)\delta$$

where T_1 and T_2 are the temperatures of the two interior surfaces and δ the gap width. In terms of Nusselt number based on gap width this becomes

$$Nu = h\delta/k = q\delta/[k(T_1 - T_2] = 1$$

so
$$Nu = 1 \tag{4.40}$$

The simplicity of this result is sufficient justification for basing the Nusselt number on the gap width rather than the plate height. Equation 4.40 is valid in the limit of low Rayleigh numbers. The Ra number is based upon δ and $(T_1 - T_2)$ as the temperature difference.

At high Rayleigh numbers a particular type of turbulence sets in (see Fig. 4.10). We could attempt to model this as two separate turbulent boundary layers, one associated with each surface, i.e. equation 4.29 would apply to each layer. Bearing in mind that equation 4.29 was based on $T_w - T_{bulk}$, and assumed a bulk temperature midway between the two interior wall temperatures, the heat transfer equation becomes

$$Nu = 0.05 \, Ra^{1/3} \tag{4.41}$$

if we base it on $T_1 - T_2$. We could also argue that the characteristic length in the Ra number should be δ, not the plate height, since this is more in keeping with the linear dimensions of the eddies (figure 4.10). In fact, in turbulent flow, since the heat transfer coefficient is independent of the value of the linear dimension, all that matters is that the same linear quantity is used in the Nu and Ra definitions.

An experimental study of heat transfer in this geometry, by ElSherbiny et al[18], did not quite agree with equation 4.41. In the limit of high Rayleigh numbers they found

$$Nu = 0.0605 \, Ra^{1/3} \tag{4.42}$$

The range of aspect ratios AR (i.e. height / gap width) covered was from 5 to 110. The enclosure was sealed at top and bottom with a conducting boundary (the horizontal dimension was considered large enough not to affect the results). At the highest aspect ratio the results are quite well represented just by equations 4.40 and 4.42. At lower aspect ratios there is a more gradual transition from one to the other. A standard method of combining two asymptotic tendencies is to write

$$Nu^n = 1 + (0.0605 \, Ra^{1/3})^n \tag{4.43}$$

for (in this case) $10^2 < Ra < Ra_{max}$

Regardless of the value of n this gives equation 4.40 in the low Ra limit and equation 4.42 in the high Ra limit. The value of n only affects the abruptness of the transition from one equation to the other.

Putting
$$n = AR^{0.45} \tag{4.44}$$

reproduces the results given by the more accurate equations[18] (a separate equation for each AR value) reasonably well. Apart from $AR = 5$ all the results

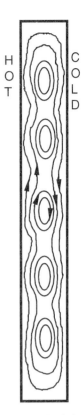

Fig. 4.10 Turbulent circulation in cavity

are filted, to within 13%. The highest Ra value tested, Ra_{max}, varied from 10^7 at $AR = 10$ to 1.2×10^4 at $AR = 10$.

Example 4.2

A 1.2 m high by 2 m wide double glazed window separates a room at 22 °C from the outside of the building at $-2°$ C. The gap between the glass panes is 12 mm. Estimate the rate of heat loss through the window. Ignore the resistance to heat transfer by conduction through the glass. Take all properties at 10 °C.

As a check find the temperature drop through each pane given a thickness of 5 mm and a thermal conductivity of 0.8 W m^{-1} K^{-1}.

The problem is simple enough to state but we need to reflect before we start on the solution. Since temperatures are given we can assume that the glass surfaces are constant temperature surfaces (certainly this is better than a constant heat flux assumption). We have equations for natural convection heat transfer for the two geometries involved—vertical plate and vertical enclosure—but they require known surface temperatures, so we will have to assume temperature values and iterate.

In simple physical terms each of the four glass surfaces has a boundary layer of slow moving fluid attached to it, i.e. there are four barriers to the heat flow. If all the barriers were equal then the temperature drop across one layer would be $(22 - [-2])/4 = 6$ K. So to start the iteration we assume that the temperature of the first glass pane, T_1, is 16 °C and that of the second, T_2, is 4 °C. With these guessed values of the temperatures we can calculate all the heat transfer coefficients, and combine them in series to find the heat flux.

To find the h values for the vertical surfaces on the inside and outside of the window the calculation is much as in example 4.1, so it is only outlined here. At this stage the symmetry of the problem means that h_{in} and h_{out} are the same.

L is 1.2 m. $T_w = 4$ °C and $T_\infty = -2$ °C. For air at $T_f = 10$ °C, $\rho = 1.248$ kgm^{-3}, $c = 1003$ J kg^{-1} K^{-1}, $k = 0.02486$ W m^{-1} K^{-1} and $\mu = 1.773$ 10^{-5} kg m^{-1} s^{-1}. $g = 9.8$ m s^{-2}, and $\beta = 1/(273 + 10) = 0.003534$ K^{-1}.

$$Ra = \frac{\rho^2 cg\beta[T_w - T_\infty]L^3}{k\mu} = 1.274 \times 10^9$$

equation 4.29 gives $Nu = 0.1 \, Ra^{1/3} = 108.4$

so $\qquad h_{out} = Nuk/L = 2.246 \qquad$ W m^{-2} K^{-1} $(= h_{in})$

To calculate the heat transfer in the gap we use equations 4.43 and 4.44, noting that Nu and Ra are now calculated with the gap width, 12 mm. The aspect ratio is $AR = 1.2/0.012 = 100$.

The physical property values are all as before but the temperature difference is 12 K.

So $Ra = 2548$ and both Ra and AR are within the valid range of the equations, n works out at 7.94 and Nu at 1.025 giving

$$h_{gap} = Nuk/L = 1.025 \times 0.024 \, 96/0.012 = 2.134 \qquad \text{W m}^{-2} \text{ K}^{-1}$$

Using the standard equation (2.11) for the heat flux through a number of layers in series,

$$q = \frac{\Delta T_{total}}{\frac{1}{h_{in}} + \frac{1}{h_{gap}} + \frac{1}{h_{out}}}$$

we find $\qquad\qquad q = 17.66 \qquad$ W m^{-2}

A better value of T_1 can now be calculated as $22 - q/h_{in} = 14.14$ °C. Similarly $T_2 = -2 + q/h_{out} = 5.86$ °C.

Repeating the calculation with these temperatures in place of the initially guessed temperatures gives

$$q = 18.62 \quad \text{W m}^{-2} \text{ first iteration)}$$

It is doubtful if further iteration is justified but if we do it anyway we find

$$q = 18.5 \text{ W m}^{-2} \text{ (second iteration)}$$

Although we have used a complicated equation to calculate the gap heat transfer we would have obtained virtually the same result if we had assumed pure conduction in the air in the gap, suggesting that the gap width could be increased to advantage.

With this heat flux value we can estimate the temperature drop through the glass pane from the Fourier equation (2.2)

$$\Delta T = q \, \Delta x / k = 18.5 \times 0.005/0.8 = 0.1 \text{ K}$$

so even allowing for the two panes the temperature drop through them is less than 1% of the total, and can be neglected.

Table 4.1   Heat transfer correlations for isothermal surfaces

Geometry	Length scale for Nu, Gr, Ra	Minimum Ra	Maximum Ra	Equation
Vertical plate	Height	10^4	10^9	4.26 and 4.28
Vertical Plate	Height	10^9	10^{13}	4.29
Wide vertical cylinder	Height			As plate*
Inclined heated plate facing down[†]	Length			4.31
Inclined heated plate facing up[†]	Length		Laminar	4.31
Inclined heated plate facing up[†]	Length		Turbulent	4.32
Horizontal heated plate facing up[†]	Shorter side	10^7	10^{11}	4.34
Horizontal heated plate facing down[†]	Shorter side	10^6	10^{11}	4.33
Horizontal cylinder	Diameter	10^4	10^7	4.35
Horizontal cylinder	Diameter	10^7	10^{12}	4.36
Horizontal Cylinder	Diameter	10^{-6}	10^9	4.39
Sphere	Diameter	1	8×10^8	4.37
Vertical enclosure	Gap width	10^2	10^7 to 10^4	4.43 and 4.44

* assuming equation 4.30 is satisfied.
[†] or for cooled surfaces facing in the opposite direction.

4.5   Spacing of fins

Often the heat transfer that is obtained by natural convection from a plane surface is insufficient to provide the required cooling. With air particularly the heat transfer coefficients are rather low, which is the condition needed to make adding fins to the surface worthwhile (as shown in Chapter 2). Vertical fins are more effective since they do not impede the flow in the boundary layer. To a good approximation the surface can still be regarded as a vertical plate as regards calculating the value of *Nu*.

At some stage though, if the fins are too close together, there will not be room for the boundary layers typical of a flat plate to form. Equation 4.21 (for $Pr = 0.7$) shows that the boundary layer thickness a vertical distance x above the lower edge of the heat transfer surface is given by:

$$\delta^4 = \frac{566\,\mu k}{\rho^2 cg\beta[T_\mathrm{w} - T_\infty]}x$$

So if the boundary layer is to develop fully the spacing s between the fins must satisfy

$$s \gg 2\delta \tag{4.45}$$

Accepting that the case of main interest is that of air at atmospheric pressure, and in addition assuming a film temperature of 40 °C and a temperature difference of 40 K, the condition on the fin spacing becomes

$$s \gg 0.042\,x^{1/4} \tag{4.46}$$

which is of limited use if we do not know how much greater the spacing has to be. Comparing these equations with the recommendations in reference 19 we find

$$s \geq 100\delta \tag{4.47}$$

and for the assumed air conditions s ≥ 6 mm.

In fact these can be regarded as optimum spacings since any spacing higher than the minimum will give a smaller number of fins and a reduced total heat transfer.

4.6   Constant heat flux surfaces

For turbulent conditions where the heat transfer correlation takes the form

$$Nu = \text{constant} \times Ra^{1/3}$$

the heat transfer coefficient is independent of distance along the surface. With constant heat flux, q, this gives a constant surface temperature, i.e. the two boundary conditions are equivalent and the equations previously suggested for turbulent conditions are all valid.

Even for laminar conditions the heat transfer coefficient does not vary that much over the surface and all of the previous equations are valid as a first approximation.

There is a practical difficulty in using the equations: the calculation of the Rayleigh number, *Ra*, requires a value for the temperature difference, which is unknown. A solution could be obtained simply by guessing the temperature difference and iteration. A better approach is to define a modified Rayleigh number (even here a guessed film temperature, usually half way up the surface, is required for property calculations).

The isothermal heat transfer correlations can generally be written

$$Nu = C \, Ra^n \qquad (4.48)$$

where C is a constant and n is typically $\frac{1}{4}$ for laminar conditions and $\frac{1}{3}$ for turbulent. In terms of the definitions of Nu and Ra this becomes

$$\frac{hL}{k} = C \left[\frac{\rho^2 cg\beta L^3 q}{k\mu} \right]^n \left[\frac{1}{h} \right]^n$$

where ΔT has been replaced by q/h. Rearranging this to make Nu the subject gives

$$\frac{hL}{k} \left[\frac{hL}{k} \right]^n = C \left[\frac{\rho^2 cg\beta L^3 q}{k\mu} \right]^n \left[\frac{L}{k} \right]^n$$

or

$$Nu^{n+1} = C \left[\frac{\rho^2 cg\beta L^4 q}{k^2 \mu} \right]^n$$

i.e.

$$Nu = C^{1/(n+1)} \, Ra^{*n/(n+1)} \qquad (4.49)$$

where the modified Rayleigh number is defined as

$$Ra^* = \frac{\rho^2 cg\beta L^4 q}{k^2 \mu} \qquad (4.50)$$

As an example we can apply equation 4.49 to the equations previously given for isothermal vertical plates. For laminar conditions equation 4.26 transforms to

$$Nu = 0.58 \, Ra^{*1/5}$$

and equation 4.29 for a turbulent boundary layer transforms to

$$Nu = 0.18 \, Ra^{*1/4}$$

These two equations are within 10% or so of equations that have been specifically recommended for constant heat flux conditions (References 20 and 21 respectively). For other geometries equation 4.49 can equally be used to transform an isothermal equation represented by equation 4.48.

References

1. H. B. Squire in *Modern Developments in Fluid Dynamics, Vol 2*, ed. S. Goldstein, Oxford, 1938.
2. O. A. Saunders, Natural convection in liquids, *Proc. Roy. Soc. A*, **172**, 55–71, 1939; also **157**, 278–291, 1936.
3. H. Schlichting, *Boundary layer theory*, (7th edn). McGraw-Hill, 1979. (reissued 1987).
4. A. Bejan and J. L. Lage, The Prandtl number effect on the transition in natural convection on a vertical surface. *Trans. ASME J. Heat Transfer* **112**, 787–790, 1990.
5. V. L. Vitharana and P. S. Lykoudis, Criteria for predicting the transition to turbulence., *Trans. ASME J. Heat Transfer* **116**, 633–638, 1994.
6. F. J. Bayley, An analysis of turbulent natural convection heats transfer. *Proc. I. Mech. E.* **169**, 361–368, 1955.
7. K. O. Pasametoglu, Turbulent natural convection to gases at high wall temperature. *Trans. ASME J. Heat Transfer* **116** 246–247, 1994.

8. S. W. Churchill and H. H. S. Chu, Correlating equations for laminar and turbulent free convection from a vertical plate, *Int. J. Heat Mass Transfer*, **18**, 1323–1331, 1975.
9. T. Fujii and H. Imura, Natural convection heat transfer from a plate with arbitrary inclination. *Int. J. Heat Mass Transfer* **15**, 755–767, 1972.
10. M. Al-Arabi and B. Sakr, Natural convection heat transfer from inclined isothermal plates. *Int. J. Heat Mass Transfer,* **31**, 559–566, 1988.
11. V. T. Morgan, The overall convective heat transfer from smooth circular cylinders. *Advances in Heat Transfer*, **11**, 199–264, 1975.
12. S. B. Clemes *et al.*, Natural convection heat transfer from long horizontal isothermal cylinders. *Trans ASME J. Heat Transfer,* **116**, 96–104, 1994.
13. T. Yuge, Experiments on heat transfer from spheres including natural and forced convection, *ASME J. Heat Transfer*, **82**, 214–220, 1960.
14. W. S. Amato and C. Tien, Free convection heat transfer from isothermal spheres in water. *Int. J. Heat Mass Transfer*, **15**, 327–339, 1972.
15. J. H. Lienhard, On the commonality of equations for natural convection from immersed bodies. *Int. J. Heat Mass Transfer*, **16**, 2121–2123, 1973.
16. E. M. Sparrow and A. J. Stretton, Natural convection from variously oriented cubes and other bodies of unity aspect ratios. *Int. J. Heat Mass Transfer*, **28**, 741–752, 1985.
17. S. W. Churchill and H. H. S. Chu, Correlating equations for laminar and turbulent free convection from a horizontal cylinder. *Int. J. Heat Mass Transfer*, **18**, 1049–1053, 1975.
18. S. M. ElSherbiny, G. D. Raithby, and K. G. T. Hollands, Heat transfer by natural convection across vertical and inclined air layers. *Trans. ASME J. Heat Transfer*, **104**, 96–102, 1982.
19. G. N. Ellison, *Thermal computations for electronic equipment.* van Nostrand Reinhold, 1984.
20. T. Fujii and M. Fujii, The dependence of local Nusselt number along a vertical surface with uniform heat flux. *Int. J. Heat Mass Transfer*, **19**, 121–122, 1976.
21. G. C. Vliet and C. K. Liu, An experimental study of turbulent natural convection boundary layers. *Trans. ASME J. Heat Transfer*, **91**, 517–524, 1969.

Problems

4.1. A 1 m high vertical surface is maintained at 40 °C. It is cooled by air at 20 °C. At what point does the boundary layer become turbulent? How thick is the boundary layer at this point? (Assume that the large Prandtl number solution is valid.)   [0.81 m from base; 18 mm]

4.2. It has been suggested that the variation in temperature T with distance y from the wall in laminar boundary layer, thickness δ, can be fitted to a polynomial

$$T = a_0 + a_1 y + a_2 y^2$$

Use the boundary conditions: $T = T_w$ at $y = 0$; $T = T_\infty$ at $y = \delta$; and a smooth merging of temperature at $y = \delta$, to find an expression for T.

4.3. Since the properties of air do not vary much over a limited range of temperature it is convenient to have an approximate equation where all the property values have already been worked out and none appear in the equation. The equation is to be valid for atmospheric pressure and $-10 \leq T_\infty \leq 30\,^\circ\text{C}$ and $10 \leq (T_w - T_\infty) \leq 90$ K.

For a film temperature of 300 K show that $Nu = C\,Ra^{1/4}$ converts to a prediction for air of

$$h = 2.58\,C\,(\Delta T/L)^{1/4} \qquad \text{W m}^{-2} \qquad\qquad \text{(A)}$$

Is this equation consistent with a recommended equation for a horizontal cylinder of $h = 1.32(\Delta T/L)^{1/4}$?   [A is 6% low]

From the point of view of property variation, what is the error in equation A for a film temperature of 325 K?   [2%]

4.4. One side of a 0.5 m high by 3 m wide vertical surface at 100 $^\circ$C is exposed to still air at 0 $^\circ$C and 1 atm. What is the rate of heat loss due to natural convection?   [723 W]

4.5. A 2 kW horizontal immersion heater is to be used to maintain the temperature of a large tank of water at 60 $^\circ$C. Boiling of the water on the surface of the heater is to be avoided so a maximum surface temperature of 90 $^\circ$C is specified. If the diameter of the heater is 50 mm what length is required? The volume expansion coefficient of water is 6.21×10^{-4} K^{-1}.   [356 mm]

4.6. A 100 mm outside diameter steam pipe has a surface temperature of 140 $^\circ$C. Calculate the rate of heat loss over a 10 m horizontal length by natural convection if the ambient air temperature is 14 $^\circ$C.   [2.80 kW]

4.7. Part of a personal computer system is essentially a box 40 cm wide by 50 cm deep by 10 cm high. 25 W of heat are generated within the box. Because of the possibility the box could be surrounded on the sides by other equipment it is assumed that heat can only be lost from the top surface. Calculate the temperature of the box if the surrounding air is at 20 $^\circ$C.   [46 $^\circ$C]

4.8. It is often stated that there is an optimum spacing for double glazing, i.e. a certain gap width gives a minimum heat loss. Is this true of the equations in this chapter?

A 1.4 m high double-glazed window separates a room at 20 $^\circ$C from the outside of the building at 0 $^\circ$C. The gap between the panes is 15 mm. Estimate the heat flux through the window. Ignore the resistance to the heat transfer by conduction through the glass. Take all properties at 10 $^\circ$C.   [13.8 W m^{-2}]

5 Radiation

5.1 Introduction

No physical medium is needed for heat transfer by radiation; energy transmission by radiation is most effective in a vacuum. The radiation is electromagnetic, the same in principle as light or radio waves.

Heat radiation is associated with objects, such as the sun or red-hot metals, that are so hot that they are also giving out visible light. There is no change in principle with cooler objects; just because the human eye cannot detect the radiation does not mean it is not still there. This radiation is known as the infrared. It carries thermal energy with it and travels at the speed of light in exactly the same way as the light rays from the sun.

All objects emit radiant energy, the amount increasing with their temperature. Some surfaces, e.g. black paint, emit more energy then others, e.g. highly polished metal. It is known that surfaces that emit more radiation tend also to absorb more of the radiation that falls upon them.

Black body

This is an important simplifying concept. In ordinary usage a black body does not reflect light. In radiation heat transfer a black body does not reflect any wavelength of radiation.

Fig. 5.1 shows how a black body could be constructed. A small hole is made in a constant temperature enclosure. Any incident radiation falling on the hole will, effectively, be absorbed. Each time it is reflected inside the enclosure more of the radiation will be absorbed and none of it escapes again through the hole. The hole behaves as a black body, absorbing all radiation that falls upon it.

Equally, though, thermal radiation exists within the enclosure, resulting from the thermal movement of the atoms. The radiation issuing from the hole is black body radiation, characteristic of the temperature of the enclosure but independent of the material of construction. This last point might seem rather surprising. If the inside of the enclosure were made of some matt, black material, it should emit more radiation than if the inside were made of a highly reflecting metal. In fact, as regards emission from the surface, the matt black surface will emit more. However the metal surface will reflect more of the radiation that is emitted and the radiation leaving the hole is the same in either case.

Fig. 5.2, with two black bodies of different construction but at the same temperature, shows that any other conclusion would be a violation of the second law of thermodynamics. If the rate at which energy leaves the first black body, made of shiny metal, was greater than the rate at which it leaves the second black body, made of matt black material, then there would be a net exchange of energy between two bodies at the same temperature. The second law states that heat can only flow from a hotter to a cooler body.

Constant temperature enclosure

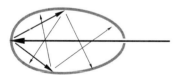

Fig. 5.1 Black body absorbs all incident radiation

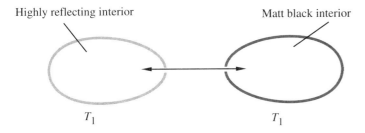

Fig. 5.2   Radiation from a black body is the same regardless of materials

A fairly obvious conclusion from this is that there must be a precise relationship between the ability of the surface to emit radiation and its ability to reflect radiation. This is proved later on (Kirchhoff's law).

Most surfaces that appear black behave roughly as black bodies. Frequently other non-metallic surfaces can be regarded as black to a first approximation for radiation heat transfer.

The Stefan–Boltzmann law
This states that the heat flux radiated from a black body is

$$q = \sigma T^4 \qquad \text{W m}^{-2} \tag{5.1}$$

where T is absolute temperature in K.

σ, the Stefan-Boltzmann constant, is 5.67×10^{-8} W m^{-2} K^{-4}. The law was originally established experimentally by Stefan and later Boltzmann proved it using thermodynamic arguments.

A complicating factor is that often the object of interest is exposed to other surfaces at different temperatures. The calculation of net energy exchange can involve difficult geometrical integrations.

We see that radiation heat transfer has little in common with conduction or convection. No medium is needed and we cannot assume that the heat flow rate is simply proportional to temperature difference. One of the few common features is that the second law of thermodynamics is always obeyed—heat always flows from a high temperature to a low temperature. We use the second law a number of times: regardless of how complicated the geometry we know that the net heat exchanged by radiation must be zero if the temperatures are the same.

Another curious feature of radiation heat transfer is that energy can travel from a hot object to a distant cooler object without necessarily warming the medium in between. Certainly this is true of radiation through air. With conduction or convection the material along the heat flow path is warmed, more so in fact than the receiving object.

Net heat exchange between a black body and its surroundings
Consider the simple case of a black body at temperature T_1 surrounded by a black enclosure at temperature T_2 (Fig. 5.3). The body can be any shape provided its surface is not concave anywhere, i.e. it cannot radiate to itself.

If the surface area of body 1 is A_1 then it will lose heat by radiation at a rate

$$A_1 \sigma T_1^4 \qquad \text{W}$$

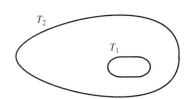

Fig. 5.3   Black body surrounded by black enclosure

from the Stefan–Bolzmann law. Similarly the enclosure, area A_2, will lose heat at a rate

$$A_2 \sigma T_2^4 \quad \text{W}$$

Suppose that the fraction of the heat from the enclosure that reaches the body 1 (and is absorbed by it, since it is black body) is F. The factor F depends purely upon the geometry. The net rate of heat loss by body 1 is

$$A_1 \sigma T_1^4 - F A_2 \sigma T_2^4$$

But when the temperatures are equal, $T_1 = T_2$, there is no heat exchange giving

$$A_1 - F A_2 = 0$$

So $F = A_1/A_2$ and the net rate of heat loss by body 1 is

$$A_1 \sigma (T_1^4 - T_2^4) \tag{5.2}$$

Example 5.1
One side of a 1 m high by 2 m wide vertical surface at 20 °C is exposed to surroundings at 10 °C. Assuming black body behaviour, what is the rate of heat loss due to radiation? (Note that these are the conditions of example 4.1 where the rate of heat loss due to natural convection was found to be 53 W.)

This is a straightforward application of equation 5.2 with $T_1 = 273 + 20 = 293$ K and $T_2 = 283$ K. The rate of heat loss is

$$2 \times 1 \times 5.67 \times 10^{-8} \times (293^4 - 283^4)$$
$$= 108 \text{ W}$$

Since the two mechanisms of heat transfer operate separately (the radiation goes through the natural convection boundary layer without warming it or affecting its structure) this result can be combined with the answer from example 4.1 to give a total rate of heat loss of $108 + 53 = 161$ W.

Example 5.2
The sun can be regarded as a black body at 5760 K with a radius of 6.96×10^5 km. The radius of the earth is 6370 km and the radius of the earth's orbit around the sun is 1.496×10^8 km. If the earth is regarded as a black body, estimate its average temperature.

The total rate at which radiant energy leaves the sun is given by equation 5.1 (multiplied by the total surface area of the sun)

$$4\pi \times (6.96 \times 10^5 \times 10^3)^2 \times 5.67 \times 10^{-8} \times 5760^4 = 3.08 \times 10^{26} \text{ W}$$

This radiation will be uniformly spread over the surface of an imaginary sphere of radius r equal to that of the earth's orbit, i.e. the heat flux is

$$3.80 \times 10^{26}/4\pi^2 r = 3.80 \times 10^{26}/4\pi \times (1.496 \times 10^8 \times 10^3)^2 = 1351 \text{ W m}^{-2}$$

(this is known as the solar constant).

The projected area of the earth towards the sun is $\pi \times (6370 \times 10^3)^2$ so the rate at which solar radiation reaches the earth (and is absorbed) is

$$1351 \times \pi \times (6370 \times 10^3)^2 = 1.723 \times 10^{17} \text{ W}$$

Assuming that the earth is losing heat by radiation from all over its surface (to space at essentially zero temperature) then the temperature T required to re-radiate this amount of heat comes again from the Stefan–Boltzmann law:

$$4\pi \times (6370 \times 10^3)^2 \times 5.67 \times 10^{-8} \times T^4 = 1.723 \times 10^{17} \text{ W}$$

giving $$T = 278 \text{ K} \qquad \text{or } 5\,^{\circ}\text{C}$$

In view of the approximations in the above calculation, the way that variations from the equator to the poles and between day and night have all been averaged out, and the assumption of black body behaviour, the answer is reasonable. A further complication is that the earth's atmosphere has a structure, with some absorption of solar radiation within the atmosphere. This gives uncertainty as to where the average temperature of 278 K refers to. Is it the surface of the earth or some point in the atmosphere? This difference is referred to again when the greenhouse effect is discussed.

A couple of points in the above analysis we will use in a more formal way later on. First, the intensity of the radiation falls off as the inverse square of the distance. In the problem there was spherical symmetry but the inverse square law applies generally for point sources or for the radiation from an element of area on a larger surface. Secondly the radiation intercepted depends on the projected area facing the source, not the total surface area.

Distribution of radiant energy with wavelength for a black body
Fig. 5.4 shows how the heat flux varies with wavelength for three typical surface temperatures. The curves are drawn using the radiation law that Planck introduced using statistical thermodynamics arguments in 1900.[1] The rate of heat emission per unit area per unit wavelength λ is

$$E_\lambda = \frac{2\pi hc^2 \lambda^{-5}}{e^{hc/\lambda k_B T} - 1} \tag{5.3}$$

h is Planck's constant (6.63×10^{-34} J s), c the speed of light 3×10^8 m s^{-1} and k_B Boltzmann's constant 1.38×10^{-23} J K^{-1}. This work was the start of what later become known as quantum theory.

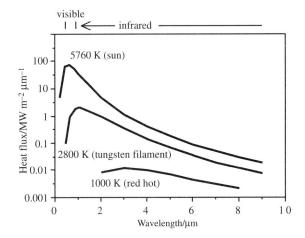

Fig. 5.4 Radiant energy versus wavelength

This work by Planck, and earlier by Boltzmann, is a remarkable example of the success of theory in heat transfer (especially after the empirical treatment of much of convective heat transfer!).

Equation 5.3 can be integrated to give the Stefan–Botzmann law:

$$\int_0^\infty E_\lambda \, d\lambda = \sigma T^4$$

Differentiating equation 5.3 is also of interest since it enables us to find the maximum of the curves in Fig. 5.4. The mathematical form of the resulting expression can be justified fairly quickly, if not the value of the constant.

If we make the substitution $x = \lambda k_\beta T / hc$ then equation 5.3 becomes

$$E_\lambda = constant \times T^5 \frac{x^{-5}}{e^{1/x} - 1}$$

At a given value of T the maximum value of this expression will occur at a certain value of x, say $x = a$, i.e.

$$\lambda_{max} k_B T / hc = a$$

or
$$\lambda_{max} T = ahc / k_B$$

and including the value of a obtained from a full analysis then

$$\lambda_{max} T = 2898 \ \mu m \ K \tag{5.4}$$

which is known as Wien's displacement law.

The same substitution, $x = \lambda k_B T / hc$, can be used to justify the T^4 dependence in the Stefan–Boltzmann law.

5.2   Real surfaces

Real surfaces display a number of features that are absent in the idealized black surface. One complication that we will not be discussing is that of specular reflection, i.e. the surface reflects like a mirror, with the angle of reflection equal to the incident angle. In this book we assume diffuse reflection: regardless of the angle of incidence the reflected radiation is spread over all angles.

The black body assumption is not too bad in practice if we avoid consideration of metals or very high temperatures, but it is possible to improve the application to real surfaces by using a factor ε_T (<1).

The total rate of heat emission per unit area is

$$\varepsilon_T \sigma T^4 \qquad W \tag{5.5}$$

ε_T is the **emissivity**.

The subscript T has been used to emphasize that the emissivity may well vary with temperature. Stictly speaking ε_T is the total hemispherical emissivity, that is it gives the integrated effect over all angles. Since in this work ε_T is the only definition of emissivity that is used it will simply be called the emissivity.

A real surface does not absorb all the radiation incident upon it. Some is reflected. With two such surfaces facing each other multiple reflections will occur, complicating the calculation of net energy exchange.

Absorptivity α and reflectivity ρ

In general, for radiation falling on a surface, there are three possibilities:

- A fraction α is absorbed
- A fraction ρ is reflected
- A fraction τ is transmitted.

with
$$\alpha + \rho + \tau = 1$$

Transmission of thermal radiation in solids is rather rare (transmission of solar radiation through glass is very much an exception). In most of this chapter we will assume $\tau = 0$.

So
$$\alpha + \rho = 1 \tag{5.6}$$

Kirchhoff's law

It turns out that there is a very simple relation between absorptivity α and emissivity ε_T.

We consider a non-black body at temperature T_1 surrounded by a black enclosure at temperature T_2 (Fig. 5.5). The analysis is similar to that to find the net heat exchange between a black body and its surroundings. In particular the same method is used to find the fraction of the energy leaving the enclosure that reaches the body 1.

If the surface area of body 1 is A_1 then it will lose heat by radiation at a rate

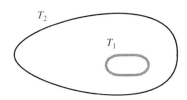

$$A_1 \varepsilon_T \sigma T_1^4$$

from the Stefan–Boltzmann law modified for a real surface. The enclosure, area A_2, will lose heat at a rate

$$A_2 \sigma T_2^4$$

The fraction of the heat from the enclosure that reaches the body 1 is F. Of this heat $FA_2 \sigma T_2^4$ reaching body 1 a fraction α is absorbed. So the net rate of heat loss by body 1 is

$$A_1 \varepsilon_T \sigma T_1^4 - \alpha F A_2 \sigma T_2^4 \tag{5.7}$$

Fig. 5.5 Non-black body surrounded by black enclosure

(α depends on both T_1 and T_2 since the nature of the radiation reaching the body depends on T_2 and the amount absorbed depends as well on T_1.)

To find the factor F we note that it is purely geometrical, unaffected by the values of ε_T or α, and so we use the same reasoning as before. Imagine that the body at 1 is replaced by a black body, with the same size, shape and position. Equation 5.7 becomes

$$A_1 \sigma T_1^4 - F A_2 \sigma T_2^4$$

and when the temperatures are equal, $T_1 = T_2$, there is no heat exchange so

$$A_1 - F A_2 = 0$$

$F = A_1/A_2$ as before and incorporating this result in equation 5.7 gives the net rate of heat loss by the original body 1 as

$$A_1 \varepsilon_T \sigma T_1^4 - \alpha A_1 \sigma T_2^4 \tag{5.8}$$

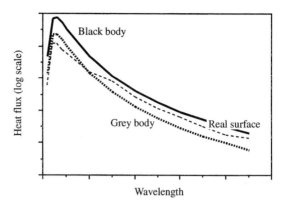

Fig. 5.6   Radiant energy versus wavelength for various surfaces

Equally though when $T_1 = T_2$ the original, non-black, body must be in thermal equilibrium with the enclosure and the net rate of heat loss is zero so

$$A_1 \varepsilon_T \sigma T_1^4 - \alpha A_1 \sigma T_1^4 = 0$$

or
$$\varepsilon_T = \alpha \tag{5.9}$$

which is Kirchhoff's law.

It is worth emphasizing that we have only proved this result, and Kirchhoff's law is only valid, for $T_1 = T_2$. For a given temperature $\varepsilon_T = \alpha$.

Also, using equation 5.6 again, $\varepsilon_T = 1 - \rho$, the relationship mentioned in the introduction.

Grey body

In this simplifying assumption a **grey body** is defined as one where the rate of heat radiation from the surface at **all** wavelengths (or temperatures) is only a fraction ε of that from a black body (Fig. 5.6).

So if we make the grey body assumption then $\varepsilon = $ constant regardless of temperature and Kirchhoff's law tells us that $\alpha = \varepsilon = $ constant regardless of temperature.

Net heat loss by grey body in a black enclosure
The analysis of the previous section gave the net rate of heat loss by a non-black body in a black enclosure directly, equation 5.8. In addition for a grey body $\varepsilon = \alpha$, independent of temperature, so the net rate of heat loss by the grey body 1 is

$$A_1 \varepsilon \sigma (T_1^4 - T_2^4) \tag{5.10}$$

This result also applies to a small grey body in a large grey enclosure. Although in principle the radiation from the grey body can now be reflected back to it, in practice this is negligible—the enclosure behaves as though it is black so far as radiation from the small grey body is concerned (Fig. 5.7).

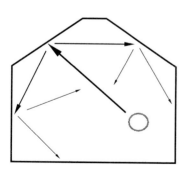

Fig. 5.7   Small grey body. Enclosure behaves as though it were black

Example 5.3
A thermocouple, emissivity 0.7, measures the temperature of combustion gases in a duct as 420 °C (Fig. 5.8). The walls of the duct are at 200 °C. The gas velocity is 1.3 m s^{-1}.

If the thermocouple may be modelled as a 1 mm diameter circular cylinder in cross-flow and the gas properties taken to be those of air at 1 atm., what is the true temperature of the gas at the thermocouple position?

The temperature of the thermocouple, 420 °C, results from a balance between radiation heat transfer to the surroundings, effectively all at 200 °C, and convection from the gas at T_g. We can regard the thermocouple as a small grey object in a large grey enclosure, so equation 5.10 applies. The heat balance becomes

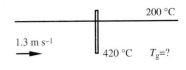

Fig. 5.8 Thermocouple of example 5.3

$$A\varepsilon\sigma(T_t^4 - T_d^4) = Ah(T_g - T_t)$$

where T_t is the temperature of the thermocouple, $273 + 420 = 693$ K and T_d is the temperature of the duct wall, $273 + 200 = 473$ K.

We need a value of the heat transfer coefficient h. The expression for cylinders in cross-flow, equation 3.24, is

$$Nu = 0.24 + \{0.4Re^{1/2} + 0.06Re^{2/3}\}Pr^{0.4}(\mu/\mu_w)^{1/4}$$

To evaluate this we need properties at the bulk gas temperature, which is unknown, although it must be higher than 420 °C. Guessing $T_g = 500\,°C$, then

$$Re = \rho u d/\mu = 0.456 \times 1.3 \times .001/3.561 \times 10^{-5} = 16.65$$

$(\mu/\mu_w)^{1/4}$ works out at 1.017 and $Pr^{0.4}$ at 0.866, giving $Nu = 2.022$. So $h = Nuk/d = 110.4$ W m^{-2} K^{-1}.

Substituting values into the heat balance equation

$$0.7 \times 5.67 \times 10^{-8}(693^4 - 473^4) = 110.4\,(T_g - 693)$$

giving $T_g = 757.9$ K $= 484.9\,°C$.

So the assumed gas temperature for the property calculations was not quite correct. Repeating with properties taken at 480 °C gives $T_g = 483.9\,°C$, not a significant change.

We see that there is a significant error in the thermocouple reading. This could be largely eliminated by a radiation shield around the thermocouple (see later).

Values of emissivity
Some values of emissivity are given in Table 5.1. A difficulty with a table like this is that emissivity is a surface property and the results for metals, particularly, depend strongly on surface condition. For this reason no attempt has been made to give the table an air of great accuracy.

Although we have stated that thermal radiation is electromagnetic radiation we have made no use of any knowledge of electricity that we might possess. It is interesting to think about the reflection of radiation from a metal with a very high electrical conductivity.

Table 5.1   Emissivities at 373 K (100 °C)

Emissivity	Type of material	Examples
0.02	Metals. Very low resistivity	Gold, silver, copper
0.04	Metals. Low resistivity	Aluminium, zinc, tungsten
0.06	Most other metals	
0.08		Chromium
0.12	Steel, stainless steel	
0.2 to 0.3	Lightly oxidized metal	Aluminium, copper, zinc
0.7 to 0.8	Heavily oxidized metal	Copper, iron, steel
0.93 to 0.98	Non-metals	Asphalt, brick, concrete, soot, water, ice, paints (non-metallic)

Note that the metal surfaces, unless otherwise specified, are assumed to be clean and polished. The information in this table has been gathered from a number of texts but the main source is reference 2.

An electric field cannot exist inside a perfectly conducting metal. Faraday felt sufficiently confident about this to climb inside a metal cage and have it charged up to the point where sparks were seen to coming from the outside of the cage. He showed that no field could be detected inside the cage. The implication of this for an electromagnetic wave being reflected from a perfectly conducting metal is that at the surface the total electric field must be zero. This is only possible if the electric field of the reflected wave is equal in amplitude (but opposite in phase) to that of the incident wave. In other words the wave is completely reflected. The reflectivity is 1 and the absorptivity (from equation 5.6) is 0. Kirchhoff's law then states that the emissivity of the surface is 0.

Real metals are not, of course, perfect electrical conductors but there are some very low ε values in the Table 5.1. More detailed analysis is possible as long as the metal can be regarded as a continuum, i.e. the electromagnetic radiation is not sufficiently energetic to be able to interact with individual electrons in the atoms. The fact that silver, copper and gold surfaces appear different colours in visible light shows that individual photon/electron interactions are occurring. The following relations cease to be valid as the visible light region is approached, i.e. as temperatures start to exceed around 1000 K.

An approximate result[3] is

$$\varepsilon_T = \text{constant} \times (\rho_e T)^{1/2} \tag{5.11}$$

for the infrared region. ρ_e is electrical resistivity. (For $\rho_e = 0$ then $\varepsilon_T = 0$, as with our simple reasoning earlier.)

Since the electrical resistivity tends to increase as temperature rises we can expect ε_T to increase with T for the clean metal surfaces. For pure metals ρ_e is quite closely proportional to T and equation 5.11 becomes

$$\varepsilon_T = \text{constant} \times T \tag{5.12}$$

So for pure metals the emissivity values in Table 5.1 can be extrapolated to higher and lower temperatures simply by assuming that ε_T is proportional to the absolute temperature. This works less well for alloys. In particular it breaks down for certain resistance wire and thermocouple materials where ρ_e is more or less constant.

Non-metallic surfaces are less affected by changes of temperature and the same ε values can be used, as a first approximation, up to around 1000 K.

The information given here will obviously only give approximate values of ε but in practice there is the problem of deciding what the surface condition is. Is one's specimen of copper to be regarded as clean and polished, lightly oxidized or heavily oxidized?

Solar radiation

The temperature of the surface of the sun is about 5760 K [4] It is unreasonable to expect that the grey body approximation will hold all the way from ambient temperatures to that of the sun, i.e. that for a given material, we can use the same emissivity and absorptivity values for radiation at the two temperatures. Table 5.2 gives the absorptivity values for solar radiation falling on various materials.

Table 5.2   Absorptivity of materials at 300 K for solar radiation

Absorptivity	Type of material	Examples
0.2	Low resistivity metals	Copper, gold, aluminium
0.2	Special white coatings	Titanium dioxide paint
0.4	Most other metals	
0.3 to 0.9	Paint	
0.7	Concrete, brick, soil	
0.9 to 0.97	Black coatings	Paint, soot
0.96		Water, ice

Note that the metals are assumed to be clean and polished. Strictly the values are normal total absorptivities, i.e. the surface should be facing the sun.

Greenhouse effect

The fact that radiation properties can be different for radiation coming from sources at different temperatures gives rise to the greenhouse effect. Transmission of the radiation is important here. In particular some materials are transparent to solar radiation but absorb infrared radiation from low temperature sources.

In the case of the original greenhouse, that is the glass structure used to house plants, sunshine travels through the glass, is absorbed by the plants and soil, and warms them. They then emit radiation characteristic of their low temperature, i.e. infrared radiation. This is absorbed by the glass. So the glass reduces the heat loss by radiation. This is only part of the story. The glass also acts as a physical barrier keeping the wind out.

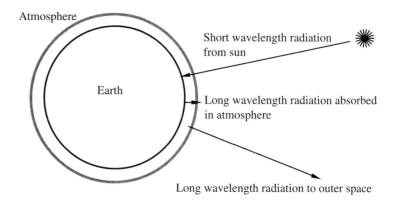

Atmosphere

Earth

Short wavelength radiation from sun

Long wavelength radiation absorbed in atmosphere

Long wavelength radiation to outer space

Fig. 5.9   Greenhouse effect

The greenhouse effect in the earth's atmosphere is illustrated in Fig. 5.9. The atmosphere is largely transparent to the sun's rays so these fall on the ground and warm it directly. The ground then re-radiates at a much longer wavelength. These longer infrared rays are partly absorbed by certain molecules in the air. In a sense the heat is trapped and thermal equilibrium is only possible if the ground temperature rises sufficiently for the total heat absorbed by the sun to be re-radiated to the sky (and then to outer space). This infrared absorption in the air tends to be associated with trace gases such as carbon dioxide, methane, and chlorofluorocarbons (CFCs). These molecules are polyatomic, as opposed to the diatomic molecules of oxygen and nitrogen. Another significant polyatomic gas, though hardly present as just a trace, is water vapour.

The natural greenhouse effect is the effect that has always been present, due to water vapour and pre-industrial concentrations of carbon dioxide. It is considered that without this natural greenhouse effect the earth's climate would be roughly 30 K colder than it actually is.[5]

The enhanced greenhouse effect is the extra warming due to artificial increases in the levels of certain gases, in particular extra carbon dioxide produced by burning fuel.

Accurate calculation of the greenhouse effect is only possible with complex numerical models. There are difficulties in the modelling, for example in accounting for the effect of clouds, and in comparing the predictions with the limited experimental data. The measured world-wide rise in temperature over the last hundred years is about 0.5 K[5] and isn't necessarily due to the enhanced greenhouse effect.[5]

In one of the problems at the end of the chapter (no. 5.7) an attempt has been made to calculate a reasonable upper limit to the total greenhouse effect. The problem talks about planet X, to emphasize that this may not be a good model of the earth, but most of the data is valid for the earth. The solar constant is correct (the rate at which solar energy is received per unit area outside the earth's atmosphere). The proportion of the solar radiation penetrating the atmosphere (about 70%) is correct for a clear day with the sun directly overhead[4]. The assumption that all of the infrared radiation emitted from the ground is absorbed by the atmosphere is wrong, and accounts for the suggestion that the calculated temperature rise (42 K) gives an approximate upper limit for the possible greenhouse effect.

A related but quite different effect is that of radiation to the sky at night. It might be thought that on a clear night radiation from the surface of the earth is travelling into outer space where the temperature is close to 0 K. This might be true for visible light but some infrared radiation (which is the relevant wavelength for ambient temperatures) is absorbed in the sky. Kirchhoff's law tells us that the sky must emit radiation appropriate to its temperature. One way of dealing with this is to pretend that the sky is a black body and define an effective sky temperature. This is usually 6 to 18 K below ground level air temperature.[4]

Net energy exchange between two grey parallel plates

The two plates are shown in Fig. 5.10. The linear dimension of the plates is assumed to be very large compared to their separation, so effectively all radiation leaving plate 1 arrives at plate 2 and vice-versa. Because these are grey surfaces some of the radiation from 1 is reflected by 2 and then some of it is re-reflected by 1 and so on indefinitely. One method of solving this problem is to sum this infinite series of reflections.

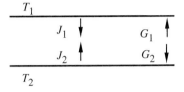

Fig. 5.10 Energy exchange between grey plates

The method that follows avoids the infinite series and introduces parameters that are useful in solving more complicated problems. Considering just surface 1 for the moment, we know that radiation will be arriving at it, due to emission and reflection from surface 2, at some rate G_1.

The rate at which radiation leaves surface 1 is

$$J_1 = \varepsilon_1 \sigma T_1^4 + \rho_1 G_1 \tag{5.13}$$

The first term represents the emission of radiation according to the Stefan–Boltzmann law, modified for a grey surface. The second term is the reflected part of G_1 that was incident on surface 1.

For the second surface we can straight away write

$$J_2 = \varepsilon_2 \sigma T_2^4 + \rho_2 G_2 \tag{5.14}$$

The radiation leaving surface 1 is the radiation approaching surface 2 and vice versa so

$$J_1 = G_2 \tag{5.15}$$

and

$$J_2 = G_1 \tag{5.16}$$

Using equations 5.15 and 5.16 to replace the G terms, equations 5.13 and 5.14 become

$$J_1 = \varepsilon_1 \sigma T_1^4 + \rho_1 J_2 \tag{5.17}$$

and

$$J_2 = \varepsilon_2 \sigma T_2^4 + \rho_2 J_1 \tag{5.18}$$

Obviously with two equations we can solve for J_1 and J_2 but we should bear in mind that finally we require the net exchange of energy, $J_1 - J_2$.

Substituting in 5.17 for J_2 gives

$$J_1(1 - \rho_1 \rho_2) = \varepsilon_1 \sigma T_1^4 + \rho_1 \varepsilon_2 \sigma T_2^4 \tag{5.19}$$

and by the symmetry of the problem

$$J_2(1 - \rho_1 \rho_2) = \varepsilon_2 \sigma T_2^4 + \rho_2 \varepsilon_1 \sigma T_1^4 \tag{5.20}$$

Subtracting 5.20 from 5.19 gives

$$(J_1 - J_2)(1 - \rho_1\rho_2) = \varepsilon_1\sigma T_1^4 - \rho_2\varepsilon_1\sigma T_1^4 + \rho_1\varepsilon_2\sigma T_2^4 - \varepsilon_2\sigma T_2^4$$
$$= (1 - \rho_2)\varepsilon_1\sigma T_1^4 - (1 - \rho_1)\varepsilon_2\sigma T_2^4$$

using equation 5.6
$$= \alpha_2\varepsilon_1\sigma T_1^4 - \alpha_1\varepsilon_2\sigma T_2^4$$

using Kirchhoff's law
$$= \varepsilon_2\varepsilon_1\sigma T_1^4 - \varepsilon_1\varepsilon_2\sigma T_2^4$$

and making the same changes for the term $\rho_1\rho_2$ on the left hand side of the equation leads to the net rate of heat exchange per unit area, $J_1 - J_2$

$$\frac{\sigma T_1^4 - \sigma T_2^4}{\frac{1}{\varepsilon_1} + \frac{1}{\varepsilon_2} - 1} \tag{5.21}$$

Radiosity

The parameters J and G introduced in the above analysis have names and we will be using them again later. J is the radiosity, the total rate at which energy leaves the surface per unit area. G is the irradiation, the total rate at which energy arrives at the surface per unit area. A curious feature of a grey surface is that energy can leave it at a much higher rate than from a black body at the same temperature. The radiosity of the grey surface can be higher if it reflects energy from a nearby hotter object.

A useful relationship for J is obtained as follows. The net rate of heat flow per unit area from either of the surfaces is

$$q = J - G$$

and if G is eliminated using equation 5.13 then

$$q = \frac{\sigma T^4 - J}{(1 - \varepsilon)/\varepsilon}$$

and including the surface area A the total rate of heat flow is

$$Q = qA = \frac{\sigma T^4 - J}{(1 - \varepsilon)/\varepsilon A} \tag{5.22}$$

An electrical analogy can be used to interpret this equation. σT^4 can be regarded as the driving force or potential on the surface, J the value of this potential immediately outside the surface, $(1 - \varepsilon)/\varepsilon A$ as a surface resistance, and q as the resulting energy flow (or current). The significance of this will be seen later when we have an expression for the extra resistance due to the distance between two radiating surfaces.

5.3   View factor

The view factor F_{12} is defined as the fraction of the energy leaving surface 1 which reaches surface 2 (Fig. 5.11). Similarly F_{21} is the fraction leaving 2 that arrives at 1.

Considering black surfaces for the moment (i.e. avoiding the complication of reflections) the rate at which energy leaves 1 and reaches 2 is

$$F_{12}A_1\sigma T_1^4$$

and the rate at which energy leaves 2 and reaches 1 is

$$F_{21}A_2\sigma T_2^4$$

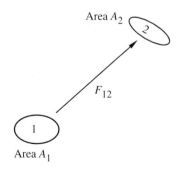

Area A_2

F_{12}

1

Area A_1

Fig. 5.11   View factor

The net rate of energy exchange (since these are black surfaces and incident energy is absorbed) is

$$F_{12}A_1\sigma T_1^4 - F_{21}A_2\sigma T_2^4$$

But when $T_1 = T_2$ the net exchange is zero so

$$F_{12}A_1 = F_{21}A_2 \tag{5.23}$$

and the net exchange, for black surfaces, can be written

$$F_{12}A_1\sigma(T_1^4 - T_2^4) = F_{21}A_2\sigma(T_1^4 - T_2^4)$$

Equation 5.23 is known as the reciprocity relation. Although we have derived it for black surfaces the view factor can be regarded as a geometric parameter and the relation is still true for grey surfaces (assuming that reflection is diffuse, i.e. spread over all angles as is the emitted radiation). Later in the chapter a general expression for evaluating the view factor is derived.

For the special case of a convex body 1 surrounded by an enclosure 2 we know that all radiation leaving 1 reaches 2, i.e. $F_{12} = 1$. Equation 5.23 shows that $F_{21} = A_1/A_2$, a result that we have already used a couple of times.

A more general result arises if the surface 1 is completely surrounded by other surfaces 2 to n, in such a way that all the radiation leaving 1 is intercepted by one or other of the remaining surfaces (Fig. 5.12). Since the total fraction of the energy leaving surface 1 that reaches the other surfaces must be 1 we have

$$F_{11} + F_{12} + F_{13} + \ldots F_{1n} = 1 \tag{5.24}$$

where F_{11} covers the possibility of surface 1 being concave and radiating to itself. For flat or convex surfaces $F_{11} = 0$.

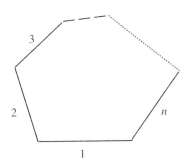

Fig. 5.12   Surface 2 to n completely surround surface 1

5.4   Heat exchange between grey bodies—electrical analogy

We assume two surfaces in series (Fig. 5.13), only exchanging radiation with one another. Since reflected radiation has to be included, the calculation is in terms of the radiosities J. Considering just surfaces 1 and 2, the total radiation leaving surface 1 per second is A_1J_1. Using the view factor the rate of energy arriving at 2 from 1 is $F_{12}A_1J_1$. Similarly the rate of energy arriving at 1 from 2 is $F_{21}A_2J_2$. The net exchange is

$$Q_{12} = F_{12}A_1J_1 - F_{21}A_2J_2$$

Using the reciprocity relation, equation 5.23, the net exchange can be written

$$Q_{12} = \frac{J_1 - J_2}{1/A_1F_{12}} \tag{5.25}$$

and the electrical analogy is extended by saying that J_1 and J_2 are the potentials (or voltages) just outside the surfaces 1 and 2 respectively and $1/A_1F_{12}$ is the resistance between these points.

Including the step from the immediate surface, equation 5.22, the complete analogous electrical network is as shown in Fig. 5.14. The logic of the analogy in this case is that we can ignore all the intermediate potentials J and write the

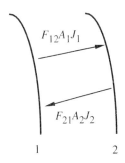

Fig. 5.13   Two grey surfaces only exchanging heat with one another

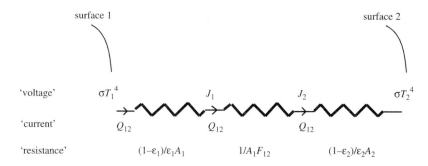

Fig. 5.14   The analogous electrical network for Fig. 5.13

net energy flow rate for a set of resistances in series in terms of the overall driving force, i.e.

$$Q = \frac{\sigma T_1^4 - \sigma T_2^4}{(1 - \varepsilon_1)/\varepsilon_1 A_1 + 1/A_1 F_{12} + (1 - \varepsilon_2)/\varepsilon_2 A_2} \qquad (5.26)$$

Extension of the network in Fig. 5.14 and of the number of terms in the denominator of equation 5.26 for several surfaces in series is straightforward. In the following section, on radiation shields, three surfaces in series are considered.

We can check equation 5.26 by applying it to some simple cases already analysed by other means. For two infinite parallel grey plates facing each other $A_1 = A_2$ and $F_{12} = 1$. Equation 5.26 reduces to the previous result, equation 5.21.

When surface 2 completely encloses body 1 equation 5.26 can be used with $F_{12} = 1$. If in addition area A_1 is very small compared to A_2 then the equation reduces to the previous result, equation 5.10.

It should be emphasized that this method, the analogy with electrical circuits, is more powerful than has been explained here. In addition to parallel and series situations it can be extended to complex networks. To give details (let alone rigorously justify the method!) requires more space than is available in this text.

Radiation shields

Sometimes the rate of heat transfer between two surfaces 1 and 2 is excessive, either in terms of heat loss from a system or because surface 2 will become dangerously hot. Another application is when body 2 is a temperature-measuring device and radiation from 1 is giving invalid readings.

To reduce the rate of heat transfer a radiation shield is placed between the surfaces (Fig. 5.15). This is a thin sheet of, preferably, low emissivity material. The emissivity of the shield material is ε_s (assumed the same on each side). The temperature drop through the shield material is assumed negligible. The analogous electrical network is shown in Fig. 5.16. It looks complicated but the intermediate potentials can be ignored. Adding up the total resistance the rate of heat flow becomes

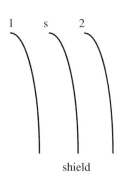

Fig. 5.15   Radiation shield placed between two surfaces

$$\frac{\sigma T_1^4 - \sigma T_2^4}{(1 - \varepsilon_1)/\varepsilon_1 A_1 + 1/A_1 F_{1s} + (1 - \varepsilon_s)/\varepsilon_s A_s + (1 - \varepsilon_s)/\varepsilon_s A_s + 1/A_s/F_{s2} + (1 - \varepsilon_2)/\varepsilon_2 A_2}$$

$$(5.27)$$

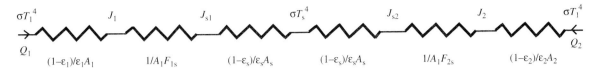

Fig. 5.16 Analogous electrical network of Fig. 5.15

If the three surfaces are long coaxial cylinders then F_{1s} and F_{s2} are both 1.

If the three surfaces are extensive parallel plates then in addition to $F_{1s} = F_{s2} = 1$ all the areas are equal. To see the effectiveness of the shield in the parallel plate case we assume that T_1 and T_2 are fixed. The heat flow rate per unit area with the shield in place is

$$q = \frac{\sigma T_1^4 - \sigma T_2^4}{1/\varepsilon_1 + 2/\varepsilon_s + 1/\varepsilon_2 - 2} \tag{5.28}$$

The heat flow rate without the shield is given by equation 5.21. If all the emissivities are the same then the shield will halve the heat flow rate. The shield is much more effective if it is made of a material with a significantly lower emissivity than surfaces 1 and 2.

5.5 General formula for view factor

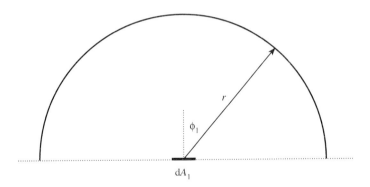

Fig. 5.17 Radiation from element of surface dA_1

First consider the way that the radiation from a small element dA_1 varies with distance and angle (Fig. 5.17). The total radiation falling on the hemisphere equals the rate at which energy is emitted from the area dA_1

$$dA_1 \sigma T_1^4$$

and the average rate at which radiation falls on unit area of the hemisphere is

$$dA_1 \sigma T_1^4 / 2\pi r^2 \tag{5.29}$$

How will the radiation vary with the angle ϕ?

$$\text{It will be maximum at } \phi = 0 \tag{5.30}$$

$$\text{It will be zero at } \phi = \pi/2 \tag{5.31}$$

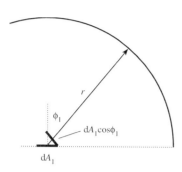

Fig. 5.18   Radiation appears to come from element $dA_1\cos\phi_1$

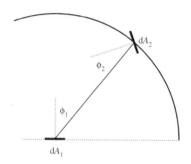

Fig. 5.19   Radiation received by element dA_2

It is found that the rate at which radiation falls on unit area is

$$\frac{dA_1\sigma T_1^4}{\pi r^2}\cos\phi_1 \tag{5.32}$$

This expression is consistent with the three equations 5.29 to 5.31. The cosine dependence is sometimes known as Lambert's cosine law. It is equivalent to saying that the intensity of the radiation is the same in all directions. If dA_1 is viewed from the angle ϕ the amount of radiation received per unit area reduces as $\cos\phi$ but equally the radiation appears to be coming from an area which is only $dA_1\cos\phi$ (Fig. 5.18).

The rate at which radiation is received by an element of area dA_2 aligned with the surface of the sphere is given by equation 5.32 multiplied by dA_2. If the normal to dA_2 is at an angle ϕ_2 to the line joining 1 and 2 (Fig. 5.19) then the projected area of dA_2 in the direction of 1 is $dA_2\cos\phi_2$ and the rate at which dA_2 receives radiation from dA_1 is

$$\frac{dA_1 dA_2\sigma T_1^4}{\pi r^2}\cos\phi_1\cos\phi_2$$

A similar argument applies to the radiation from 2 received by 1 so the net rate of energy exchange is

$$\frac{dA_1 dA_2\cos\phi_1\cos\phi_2\sigma}{\pi r^2}(T_1^4 - T_1^4) \tag{5.33}$$

Using the definition of view factor

$$F_{12}A_1 = F_{21}A_2 = \iint\frac{\cos\phi_1\cos\phi_2}{\pi r^2}dA_1 dA_2 \tag{5.34}$$

Even for simple geometries the double integral in this expression is not easy to evaluate. Although the figures in this section have been drawn two dimensional to make them easier to understand, the areas A_1 and A_2 extend in three dimensions. Results for view factor for a number of geometries are given in References 2 and 6.

Although view factors, as a purely geometric property, can be accurately calculated, their use in finding the net radiant energy exchange assumes that each surface is at a constant temperature. This may not always be an accurate assumption.

If A_1 and A_2 are both small compared to r^2 then ϕ_1, ϕ_2 and r could all be approximated by their average values, regarded as constant and taken out of the integral, i.e.

$$F_{12}A_1 = F_{21}A_2 = \frac{\cos\phi_1\cos\phi_2}{\pi r^2}A_1 A_2 \tag{5.35}$$

but this can only be used for rough estimates.

For the case mentioned before, where surface 2 completely surrounds surface 1, $F_{12} = 1$ and the reciprocity relation, equation 5.23, gives $F_{21} = A_1/A_2$.

For a few other cases, with a small number of surfaces and a high degree of symmetry, the reciprocity relation and equation 5.24 can be used to avoid the double integrations (see problem 5.8).

References

1. M. Planck, *Verhandlung der Deutschen Physikalischen Gesellschaft*, **2**, 237–245, 1900 (reprinted and translated in: *Planck's original papers in quantum physics*, Taylor and Francis, 1972).
2. R. Siegel and J. R. Howell, *Thermal radiation heat transfer*. Hemisphere, 1992.
3. R. W. Ditchburn, *Light*. Blackie, 1952.
4. L. F. Jesch, *Solar Energy Today*, International Solar Energy Society, 1981.
5. D. J. Wuebbles and J. Edmonds, *Primer on Greenhouse Gases*, Lewis Publishers, 1991.
6. H. R. N. Jones, *Radiation Heat Transfer*. Oxford University Press, 1997. In Press.

Problems

5.1. A blackbody wire is heated to 2300 K. If the total power radiated is 100 W and the diameter of the wire is 0.1 mm, what is the length of the wire?

[20 m]

5.2. A 100 mm outside diameter steam pipe has a surface temperature of 140 °C. Calculate the rate of heat loss by radiation over a 10 m length. Emissivity is 0.7 and the surroundings are at 14 °C. (Note that these are also the conditions of problem 6 in the previous chapter.) [2.78 kW]

5.3. A horizontal, square, metal plate, of 1 m side, is exposed to solar radiation of 700 W. The bottom surface of the plate is well insulated. The absorptivity for solar radiation is 0.4 and the emissivity is 0.3. The surroundings are all at 20 °C. Calculate the equilibrium temperature of the plate a) neglecting and b) including the effect of natural convection. [a) 119.9 °C b) 56 °C]

5.4. A power plant for a space probe needs to emit heat to space. In a proposed design, instead of circulating the working fluid of the power cycle through a heat exchanger, the liquid metal will be exposed directly in space. A 1 m wide sheet of liquid, 1 mm thick, is sent from the generator, at 0.01 m s^{-1}, to the collector, which is 3 m away. The liquid leaves the generator at 400 K. How much heat can be emitted and what is the temperature of the liquid as it enters the collector? [348 W, 312.8 K]

The emissivity of the liquid metal is 0.3 and its thermal conductivity is 40 W m^{-1} K^{-1}. Density and specific heat capacity are 950 kg m^{-3} and 1400 J kg^{-1} K^{-1} respectively.

5.5. Calculate the heat exchange by radiation in the cavity between the two brick walls of a house. The temperatures of the opposing surfaces are 12 and 6 °C. The emissivity of brick is 0.9.

Assuming that the temperatures of the brick surfaces do not change, what is the new rate of heat exchange if a thin aluminium sheet, of emissivity 0.1, is placed in the gap? [25.0, 1.50 W m^{-2}]

5.6. Two small surfaces. δA_1 of 10 cm^2 and δA_2 of 20 cm^2, face one another over a distance of 2 m. Estimate the view factors between the surfaces. What are the new view factors if the normal to surface 1 becomes inclined at 30° to the line joining the surfaces and surface 2 at 60°?

$$[1.59 \times 10^{-4} \text{ and } 7.96 \times 10^{-5}; 6.84 \times 10^{-5} \text{ and } 3.42 \times 10^{-5}]$$

5.7. Planet X orbits its sun at a distance of 1.496×10^8 km. The sun can be regarded as a black body at 5760 K with a radius of 6.96×10^5 km.

The atmosphere of the planet may be considered transparent to radiation apart from a layer some distance from the planet's surface that has the following properties. It reflects 10% and absorbs 20% of the incoming short wavelength radiation from the sun. It absorbs 100% of the long wavelength (infrared) radiation emitted from the planet. The average absorptivity of the planet's surface (for solar radiation) is 0.9 and the emissivity at low temperatures is 0.95.

Estimate the temperature of the absorbing layer in the atmosphere and the temperature of the planet's surface. You may assume that the planet rotates sufficiently quickly for its surface temperature to be similar on the bright and dark sides.   [266 and 308 K]

5.8. Consider the radiation exchange between the internal surfaces of long ducts of simple cross-section, i.e. effectively infinitely long prisms. By working progressively through the following, view factors can be calculated using just symmetry, the reciprocity relation and equation 5.24.

a) Cross-section is an isosceles triangle. Calculate the view factor between the base and one of the equal sides.   [0.5]

b) Cross-section is square. Calculate the view factor between adjacent sides.

[1- 1/2]

c) Cross-section is square. Calculate the view factor between opposite sides.

[0.414]

Appendix

Air properties. Values valid over a wide range of pressure (apart from density)

Temperature /°C	Viscosity /kg m^{-1} s^{-1}	Thermal Conductivity /W m^{-1} K^{-1}	Specific Heat /J kg^{-1} K^{-1}	Prandtl no.	Density at 1 atm /kg m^{-3}
−40	1.519×10^{-5}	0.021 02	1002	0.724	1.513
−20	1.624×10^{-5}	0.022 58	1002	0.720	1.394
0	1.724×10^{-5}	0.024 11	1002	0.717	1.292
20	1.822×10^{-5}	0.025 60	1003	0.713	1.204
40	1.916×10^{-5}	0.027 07	1003	0.710	1.127
60	2.008×10^{-5}	0.028 51	1004	0.707	1.059
80	2.097×10^{-5}	0.029 92	1005	0.705	0.999
100	2.183×10^{-5}	0.031 31	1007	0.702	0.946
120	2.267×10^{-5}	0.032 68	1009	0.700	0.897
140	2.348×10^{-5}	0.034 02	1012	0.698	0.854
160	2.428×10^{-5}	0.035 35	1014	0.697	0.815
180	2.506×10^{-5}	0.036 66	1017	0.696	0.779
200	2.582×10^{-5}	0.037 95	1021	0.695	0.746
220	2.656×10^{-5}	0.039 22	1025	0.694	0.715
240	2.729×10^{-5}	0.040 48	1029	0.693	0.688
260	2.800×10^{-5}	0.041 72	1033	0.693	0.662
280	2.870×10^{-5}	0.042 95	1037	0.693	0.638
300	2.938×10^{-5}	0.044 17	1042	0.693	0.616
320	3.005×10^{-5}	0.045 37	1047	0.693	0.595
340	3.071×10^{-5}	0.046 56	1052	0.694	0.575
360	3.136×10^{-5}	0.047 74	1057	0.694	0.557
380	3.199×10^{-5}	0.048 90	1062	0.694	0.540
400	3.262×10^{-5}	0.050 06	1066	0.695	0.524
420	3.324×10^{-5}	0.051 20	1071	0.695	0.509
440	3.384×10^{-5}	0.052 34	1076	0.696	0.495
460	3.444×10^{-5}	0.053 47	1081	0.696	0.481
480	3.503×10^{-5}	0.054 58	1086	0.697	0.468
500	3.561×10^{-5}	0.055 69	1091	0.697	0.456
520	3.618×10^{-5}	0.056 79	1096	0.698	0.445
540	3.674×10^{-5}	0.057 88	1100	0.698	0.434
560	3.730×10^{-5}	0.058 96	1105	0.699	0.423
580	3.785×10^{-5}	0.060 03	1109	0.699	0.414
600	3.839×10^{-5}	0.061 10	1113	0.700	0.404

Water properties

Temp.	Vapour pressure	Viscosity	Thermal conductivity	Specific Heat capacity	Prandtl no.	Density
/°C	/bar	/kg m^{-1} s^{-1}	/W m^{-1} K^{-1}	J kg^{-1} K^{-1}		kg m^{-3}
0	0.006 10	1.702×10^{-3}	0.5671	4227	12.69	1001
5	0.008 72	1.484×10^{-3}	0.5756	4215	10.86	1000
10	0.012 27	1.302×10^{-3}	0.5838	4205	9.38	999
15	0.017 05	1.150×10^{-3}	0.5916	4196	8.16	998
20	0.023 39	1.021×10^{-3}	0.5991	4190	7.14	996
25	0.031 70	9.125×10^{-4}	0.6063	4185	6.30	995
30	0.042 48	8.196×10^{-4}	0.6132	4181	5.59	994
35	0.056 30	7.399×10^{-4}	0.6197	4178	4.99	992
40	0.073 86	6.711×10^{-4}	0.6259	4177	4.48	991
45	0.095 96	6.115×10^{-4}	0.6318	4176	4.04	990
50	0.1235	5.596×10^{-4}	0.6374	4177	3.67	988
55	0.1576	5.141×10^{-4}	0.6427	4178	3.34	987
60	0.1994	4.742×10^{-4}	0.6477	4180	3.06	985
65	0.2503	4.390×10^{-4}	0.6524	4183	2.82	982
70	0.3118	4.078×10^{-4}	0.6567	4187	2.60	979
75	0.3856	3.801×10^{-4}	0.6607	4191	2.41	975
80	0.4737	3.554×10^{-4}	0.6645	4196	2.24	972
85	0.5780	3.333×10^{-4}	0.6679	4201	2.10	968
100	1.013	2.795×10^{-4}	0.6764	4219	1.74	958
110	1.431	2.517×10^{-4}	0.6805	4234	1.57	951
120	1.983	2.287×10^{-4}	0.6835	4250	1.42	943
130	2.697	2.094×10^{-4}	0.6853	4268	1.30	936
140	3.607	1.932×10^{-4}	0.6859	4288	1.21	932
150	4.751	1.795×10^{-4}	0.6853	4311	1.13	921
160	6.168	1.677×10^{-4}	0.6837	4336	1.06	910
170	7.904	1.575×10^{-4}	0.6808	4365	1.01	899
180	10.01	1.487×10^{-4}	0.6769	4399	0.97	887
190	12.52	1.409×10^{-4}	0.6719	4438	0.93	875
200	15.52	1.340×10^{-4}	0.6657	4484	0.90	862
220	23.15	1.223×10^{-4}	0.6502	4598	0.86	837
240	33.42	1.125×10^{-4}	0.6304	4754	0.85	809
260	46.89	1.037×10^{-4}	0.6064	4962	0.85	779
280	64.17	9.537×10^{-5}	0.5784	5239	0.86	747
300	85.95	8.695×10^{-5}	0.5463	5600	0.89	710
320	113.01	7.799×10^{-5}	0.5104	6065	0.93	668

The error function

x	erf x	x	erf x
0	0	1.2	0.9103
0.1	0.1125	1.4	0.9523
0.2	0.2227	1.6	0.9763
0.3	0.3286	1.8	0.9891
0.4	0.4284	2.0	0.9953
0.5	0.5205	2.2	0.9981
0.6	0.6039	2.4	0.9993
0.7	0.6778	2.6	0.9998
0.8	0.7421	2.8	0.9999
0.9	0.7969	3.0	1.0000
1.0	0.8427		

Error function simplified from reference 6 of Chapter 2. Linear interpolation can be used to much better than 1% accuracy.

Property values calculated from the equations in T. E. Daubert and R. P. Donner, *Physical and Thermodynamic Properties of Pure Chemicals part 1*. Hemisphere, 1989, apart from air density which is given by the perfect gas law. With p in bar (10^5 N m^{-2}) air density $= 348.2p/T$.

Index